Computer Modeling in the Aerospace Industry

Scrivener Publishing
100 Cummings Center, Suite 541J
Beverly, MA 01915-6106

Publishers at Scrivener
Martin Scrivener (martin@scrivenerpublishing.com)
Phillip Carmical (pcarmical@scrivenerpublishing.com)

Computer Modeling in the Aerospace Industry

Edited by

Iftikhar B. Abbasov

Scrivener Publishing

WILEY

This edition first published 2020 by John Wiley & Sons, Inc., 111 River Street, Hoboken, NJ 07030, USA and Scrivener Publishing LLC, 100 Cummings Center, Suite 541J, Beverly, MA 01915, USA

For more information about Scrivener publications please visit www.scrivenerpublishing.com.

Wiley Global Headquarters
111 River Street, Hoboken, NJ 07030, USA

For details of our global editorial offices, customer services, and more information about Wiley products visit us at www.wiley.com.

Library of Congress Cataloging-in-Publication Data

ISBN 978-1-119-66131-3

Cover image: Chromakinetic | Dreamstime.com
Cover design by Kris Hackerott

Set in size of 11pt and Minion Pro by Manila Typesetting Company, Makati, Philippines

10 9 8 7 6 5 4 3 2 1

Dedicated to the 70-year anniversary of the Department of Engineering Graphics and Computer Design

Contents

Abstract

This book is devoted to unique developments in the field of computer modeling of aerospace engineering. The book describes the original conceptual models of amphibious aircraft, ground-effect vehicles, hydrofoil vessels, from the idea to the full implementation. The developed models are presented with the design of passenger compartments and are actually ready for implementation in the aircraft industry. The originality of the concepts bases on biological prototypes, they are ergonomic, multifunctional and beautiful. The aerodynamic layout of prospective convertible land and ship-based aircrafts of vertical and short takeoff-landing is presented. The development of the original model of the unmanned aerial vehicle embodied in the material is described; the results of full-scale experiments are presented. The technology of modeling aerospace simulators based on the virtual reality environment with technical vision devices is considered.

The book is intended for researchers and developers in the area of aerospace industry, for aircraft designers and engineering students.

Preface

In today's world it is difficult to imagine a person's life without business trips or travelling. Aircraft are often used for this purpose, which is not a surprise. The beauty of the surrounding dark blue sky and the clouds floating below can hardly leave anyone indifferent at an altitude of ten thousand kilometers. Outside the window, the weather is not very changeable, while there may be a thunderstorm, rain or snow on the ground. Aviation attracts not only with high speed and speed of flights, it is used in many areas of daily life. Due to the high maneuverability, versatility, it can be used for civilian, military purposes, for rescue operations, and in extinguishing forest fires. It is for these purposes that amphibious aircraft are the most effective; they are also indispensable for search and rescue operations at sea. In many coastal countries, coastal hydroaviation is very important; in island countries it is simply irreplaceable. Therefore, the design of hydroaviation and amphibious aircraft remain relevant.

The first Chapter of this book is devoted to the use of computer modeling in aircraft construction. The process of designing a conceptually new aircraft "Chiroptera" based on bionic forms is described. Also the conceptual design of the new "Lotos" car is considered. When creating the concept of the car body design, bionic shape of lotus petals was used. For clarity, the scenes of toned three-dimensional models of simulated objects are rendered.

The second Chapter presents a three-dimensional modeling of existing amphibious aircraft Be-200, Be-103, as well as "Lapwing" concept, "Water Strider" ground-effect vehicle, and "Afalina" hydrofoil vessel. The process of designing passenger cabins of aircraft is described. The second Chapter also considers computer simulation of "Sesarma" autonomous mobile robotic system. The renderer is selected and the light sources are installed and the final scenes of realistic rendering are shown.

The third Chapter discusses the various types of aerodynamic layouts and possible applications of modern and advanced convertible land and ship-based aircrafts of vertical and shortened takeoff and landing.

The concepts and requirements for convertible high-speed aircrafts with cryogenic, hybrid diesel-electric and turboprop power units are presented.

The fourth Chapter presents a conceptual model of a multifunctional amphibious aircraft. Brief technical characteristics, visual range and historical aspects of the aircraft creation are presented. The prototype of the amphibious aircraft concept was "Consolidated PBY-5A Catalina", a modification of the amphibious aircraft of the fifties. In addition to the use for travel and outdoor activities, this aircraft can be used in many other utilitarian purposes.

The fifth Chapter discusses the features of the formation of a mathematical model of an unmanned aerial vehicle with an elliptical wing. In order to create a mathematical model of object control, a number of full-scale models of the unmanned vehicle were created. The results of flight tests of large-scale models of the device are given, characteristic features of management are outlined.

The sixth Chapter discusses the technology of planning and implementation of geometric support of simulator-modeling computer systems that operate on the basis of virtual reality. The basic elements of rational discretization of complex technical objects described by curves and surfaces are given. The software architecture of the simulator-modeling complex with channels of rendering of the virtual environment and technical vision is described.

Prof. Iftikhar B. Abbasov
Editor

1

Computer Simulation in Aircraft

Iftikhar B. Abbasov

Southern Federal University, Engineering Technological Academy, Department of Engineering Graphics and Computer Design, Taganrog, Russia

Abstract

The first chapter is devoted to the use of computer simulation in aircraft construction. The process of designing a conceptually new aircraft based on bionic forms is described. Presents search visual and graphical solutions of the developed model. Three-dimensional modeling of the structural parts of the aircraft was carried out according to draft designs. To simulate three-dimensional surfaces, the Koons method is used. The issues of solid-state modeling of a space rocket using the lofting method and the Be-200 amphibious aircraft were also considered. Also considered is the conceptual design of the new car. When creating the concept of a car body design, a bionic shape of lotus petals is used. Prototypes of car body contours are designed as draft designs. For three-dimensional modeling of structural parts of the car, polygonal extrusion method was applied. A three-dimensional computer model of the developed car concept is presented. For clarity, scenes are presented rendering of shaded three-dimensional models of simulated objects.

Keywords: Conceptual model, flight of bats, three-dimensional modelling, aircraft, car, bionics

1.1 Simulation of Aircraft

To simulate a certain aircraft [1, 34], the graphic system of three-dimensional modeling Mechanical Desktop is used. In this system, it is possible to carry out modeling of both technical objects and elements of the interior or landscape. To create three-dimensional models, the following commands are

Email: iftikhar_abbasov@mail.ru

Iftikhar B. Abbasov (ed.) Computer Modeling in the Aerospace Industry, (1–22)

used: _extrude_ (extrusion), _revolve_ (rotation), _sweep_ (bending) and _loft_ (lofting). The simplest and most obvious of the 3D commands is the _extrude_ command. To create a three-dimensional body, a profile sketch is extruded along the third axis for a distance equal to a given body thickness. The profile can be extruded in any direction or in two directions at once (the so-called extrusion from the middle plane). The transformation of the sketch into a three-dimensional model is carried out after some restrictions are imposed on it. The process of imposing restrictions includes creating a profile or imposing dimensional restrictions.

Creating a model consists of several stages:

- at the first stage, we determine the sequence of creation of structural parts and the corresponding methods for this;
- at the second stage we carry out the general assembly and completion of the device as a whole;
- at the third stage, to create a realistic model, we perform toning and final rendering of the model of the aircraft.

At the first stage, we will create an aircraft wing in the form of an oblique wedge surface. To do this, we draw profiles of the wing on the working planes, and the creation of the wing itself is carried out by the loft method (the method of supporting sections). In the three-dimensional mechanical modeling system Mechanical Desktop, the loft method is implemented using the _loft_ command. First we create two working planes and build wing profiles on them, then we translate each figure into a profile, that is, we impose restrictions. After selecting the _loft_ command, we indicate the wing profiles in succession to the request. We create the second wing using the mirror reflection of the built wing and place them along the future fuselage.

At the second stage, to create the fuselage, we again select the same method of support sections. In this case, the support sections can play the role of guide frames. The result of the construction of the fuselage support sections (four) with wings is presented in the form of a frame model in Fig. 1.1.

Using the _loft_ command, we build the fuselage surface, and to create a jet engine nozzle, we use the _extrude_ command using the subtraction method. Then, to create realistic scenes of a three-dimensional model, we assign textures and light sources. Next we select the background for the scene as a photograph of the Earth from orbit. The rendering scene is shown in Fig. 1.2. It should be noted that this scene can be modeled in other three-dimensional systems; however, this graphic system allows you to make the design documentation of the object.

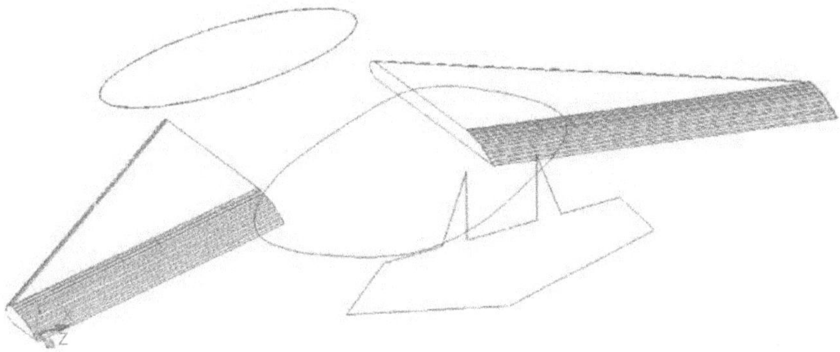

Fig. 1.1 The three-dimensional wireframe model aircraft.

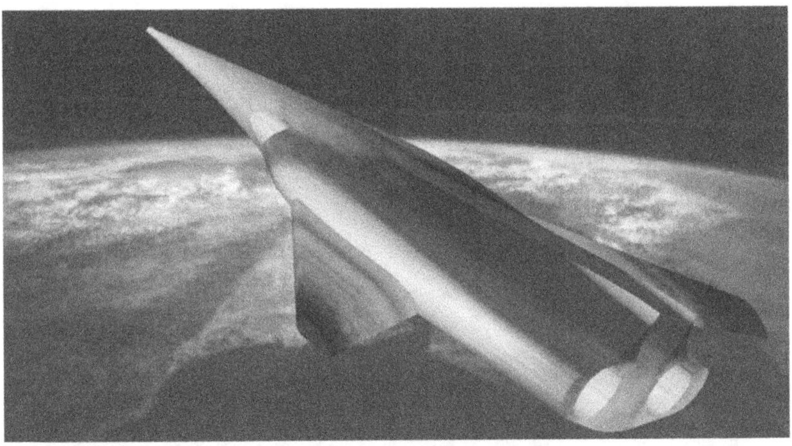

Fig. 1.2 Scene of rendering of the aircraft from the Earth's orbit.

1.2 Simulation of Rocket

Let's create a solid model of a rocket using the lofting method [35, 36] in the AutoCAD graphic system. To use the loft method, you must create the initial path of the loft and the cross-section in the form of flat figures. In the top view, with the help of two-dimensional primitives, it is necessary to construct rocket cross-sectional figures; in the front view, the axis of the rocket height is constructed. Further, the cross-section figures will be mixed to the necessary heights, as in Fig. 1.3.

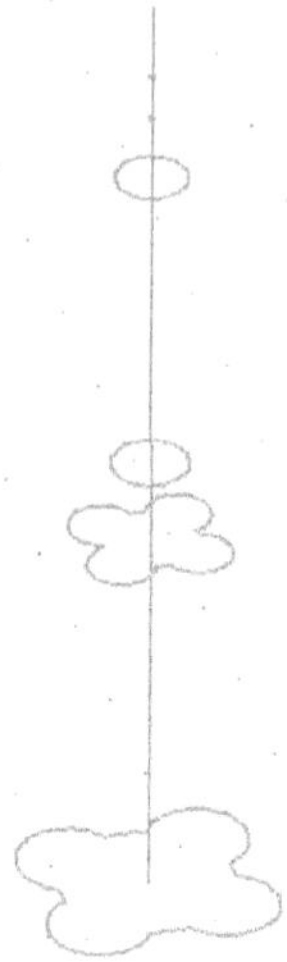

Fig. 1.3 Plane figures for lofting.

In the three-dimensional view, using the Solid\Loft command, the section shapes in height are alternately selected, and the appropriate anti-aliasing mode is selected from the Loft Setting dialog box. Fig. 1.4 shows the rendering of a shaded model of a rocket created by the loft method against the background of the Earth.

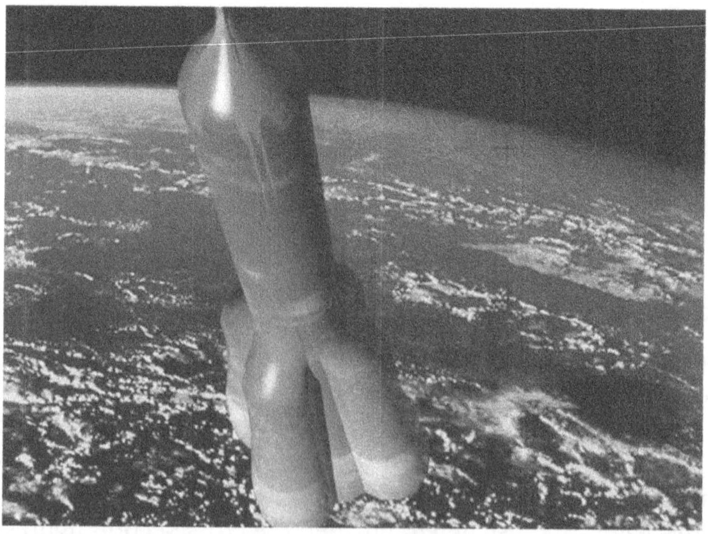

Fig. 1.4 Rendering of a shaded solid model of a rocket.

1.3 Modeling of Streamlined Surfaces

With the help of the graphic system Mechanical Desktop, you can model objects with streamlined surfaces. As such objects, their natural counterparts are most often used, and in our case a streamlined underwater object in the form of a stingray (the so-called "black devil") will be modeled [37].

To simulate the surface of the "black devil", the method of creating surfaces with several generators was used (Fig. 1.5). In the three-dimensional system of Mechanical Desktop with the help of the _edgesurf_ command, you can build unusual surfaces defined by four generating objects. Segments, arcs, splines, and polylines can act as generators. By the command _edgesurf_ a polygonal network is created – the surface of Coons, i.e., surface defined by four faces.

The image of the stingray in the form of a flat spline was inscribed in the dimensional rectangle. Then a dimensional prism was created on the basis of a rectangle; the prism in turn was divided into smaller prisms. In accordance with the image of the stingray, three-dimensional splines were inscribed into these prisms. The body of the slope was divided into several sections, and the wings were mainly built on the whole three-dimensional splines. As a result, a complete object was created by connecting different fragments.

Further, material was assigned to create realistic scenes of a three-dimensional model, and light sources were adjusted. Blue chrome was chosen as the basis for the material of the surface of the stingray. Taking into account the specific environmental conditions, the main color, shadows, color of reflection, smoothness of the surface, transparency were corrected. For the reflected color and for the unevenness of the surface, an additional text. map from a separate image was used. After assigning the material, the background was selected as a photograph of an underwater landscape. The rendering scene is shown in Fig. 1.6.

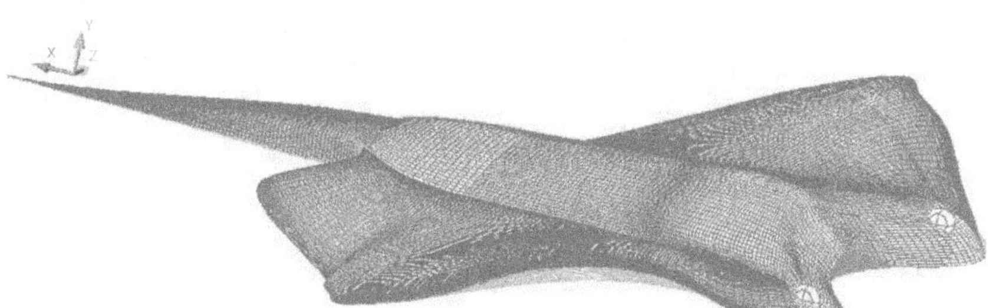

Fig. 1.5 Three-dimensional wireframe model of "Black Devil".

Fig. 1.6 Rendering of scene with the "Black Devil".

1.4 Simulation of the Be-200 Amphibious Aircraft

The issues of three-dimensional modeling of the Be-200 amphibious aircraft [2], developed by the Beriev Taganrog Aviation Scientific-Technical Complex are considered. AutoCAD graphical system is used for simulation. The three-dimensional model of an amphibian can be created using solid modeling. This will require some source material in the form of dimensional drawings (Fig. 1.7) and photos of the object being modeled (Fig. 1.8).

To model the aircraft, we divide it into its component parts: fuselage (boat), wing, tail, engines, stabilizers. When modeling, the loft method is

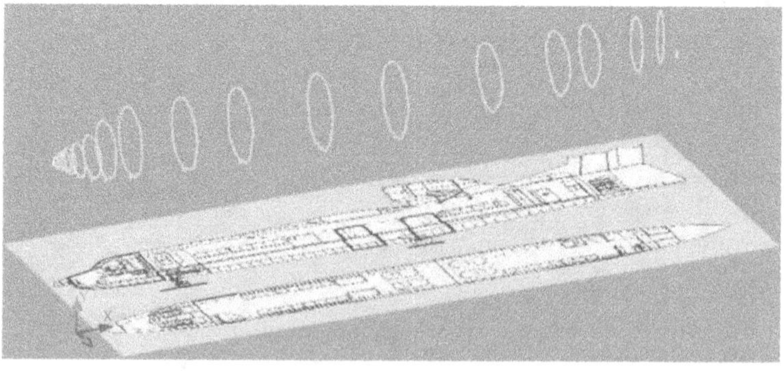

Fig. 1.7 Flat boat shapes based on drawings.

Fig. 1.8 Takeoff of the amphibious aircraft Be-200 from the water.

mainly used. For technological reasons, this is the most suitable method used in both aircraft manufacturing and shipbuilding. Fuselage frames are used as flat sections; the accuracy of the model will depend on their number.

Flat sections can be built on the basis of the spline, and it is necessary to distribute them along the length of the boat. The wing and the tail section are constructed in a similar way. Engines and stabilizers are created by rotating out of a polyline. Fig. 1.9 shows the result of modeling the structural parts of an amphibian based on the original flat forms.

Fig. 1.10 presents a theoretical drawing of an amphibian aircraft, which was automatically built on the basis of the parametric model created. For

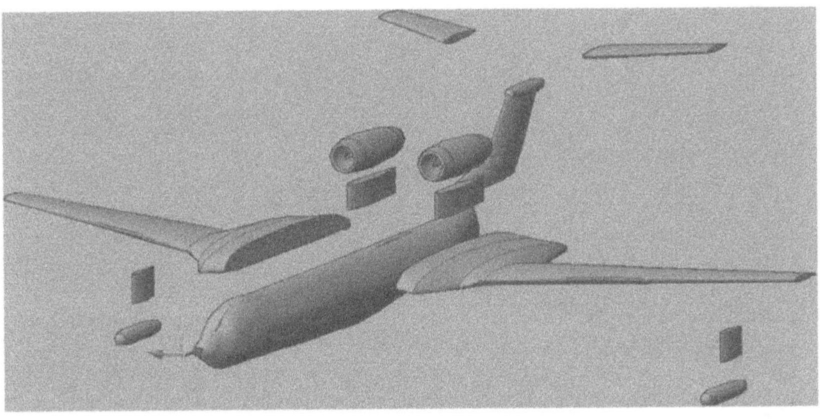

Fig. 1.9 Structural parts of an amphibious aircraft.

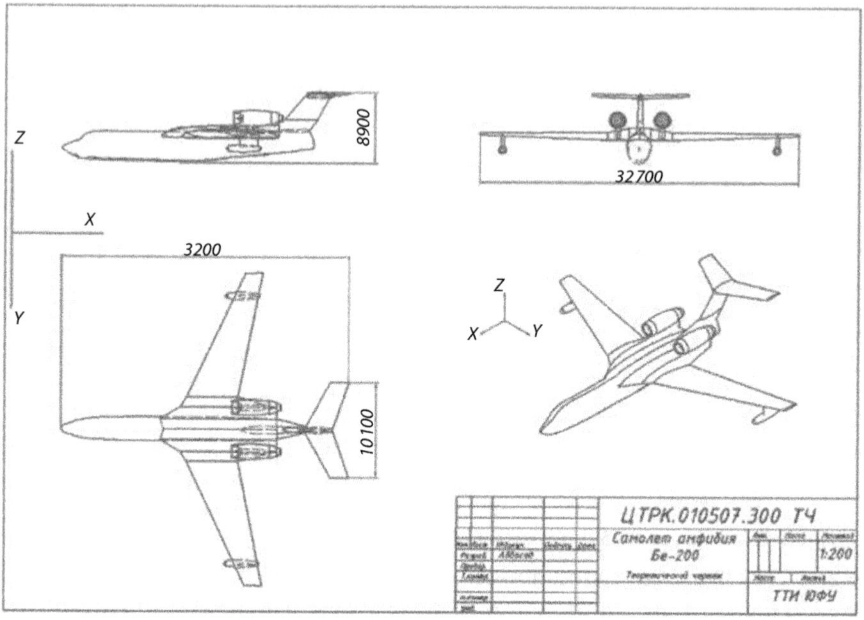

Fig. 1.10 Theoretical drawing based on parametrical model.

the development of this design document the graphic system Mechanical Desktop (Autodesk Inventor) was used, which is a three-dimensional add-on of the AutoCAD system.

Fig. 1.11 shows a solid model of an amphibian in the Realistic display mode. Fig. 1.12 presents the rendering of the scene of the amphibian.

Fig. 1.11 Solid shaded amphibian model.

Fig. 1.12 Rendering of a scene from an amphibian aircraft.

1.5 Conceptual Model of Aircraft "Chiroptera"

Day by day, the necessity of visual demonstration of industrial products under development in different spheres of human activity has been increasing. It is justified from an economic point of view. State-of-the-art technologies of computer modeling and designing enable us to achieve results which engineers could not even dream of a couple of decades ago [3].

Issues of aircraft state-of-the-art technologies of computer modeling and design application are actual. Let's consider some of the modern literature references in this sphere. The article [4] describes specific features of modern software tools implementation for design objectives. New software possibilities of aircraft structure development are described.

The article [5] is concerned with aircraft conceptual design where birds wing aerodynamic properties are taken into account. In papers [6, 7] the design of economical-to-operate passenger aircraft issues are considered.

The article [8] is concerned with conceptual design of passenger tailless airplane. Different aerodynamic configurations are presented and analyzed. In the paper [9] the conceptual design properties of new generation supersonic aircraft with original arrangement of landing gears and fuel tank are given.

This paper is concerned with 3D computer modeling of new aircraft concept. For aircraft modeling the Koons surfaces method (or Edge Mesh) is used. This method enables the creation of curvilinear streamlined surfaces based on four closed arbitrary 3D splines. Based on analysis of air environment natural forms renfering imagery, the aerodynamic characteristics of the only flying mammals – chiroptera seemed to be attractive. Skin stretched between their fingers looks like Koons surfaces built per Edge Mesh.

Wing-handed animals (*Chiroptera*) – from Greek language χείρ – *cheir* (hand), πτερόν – *pteron* (wing) [10]. Chiroptera is the only placental mammal able for active flight. They are divided into two suborders: fruit bats and bats. They are widely spread geographically (except tundra and polar regions). Their main way of travel is a flapping flight. It enables them to use biological resources inaccessible for other mammals. In general, chiroptera feed on insects, small vertebrata, fish and fruits as well. They are active in evening shades and nights; in daylight time they hide in different shelters. That's why in many cultures chiroptera, especially bats, have a bad reputation and are associated with evil magic and witchcraft.

Speed of their flight depends on wing shape [11]. Relatively high speeds are typical for lunkers who have long arrow-headed wings. Flapping flight for chiropteras is typical but some representatives use hovering and gliding flight. In flight at high speeds the birds can change their wings' length and square and chiropteras can't bend wing without all membrane relcasing. However, in contrast to birds they can freely change wing profile outline.

In the history of world aviation the first attempts to create aircraft were connected with imitation of bird flight. The first time aircraft with a wing like bats was created was by French engineer Clément Ader (1841–1925) [12]. His aerostatic machine was a tailless monoplane without vertical fin. Fuselage and wing frames were manufactured of bamboo silk-fabric covered. The aircraft was equipped with two pusher propellers with four blades, each of which was fitted with a steam engine. He called his aircraft "Eole" in honor of the ancient Greek Olympian Eolus (Fig. 1.13). In October 1890, Ader tried to fly on Eole, which took off and flew about 50 m.

Fig. 1.13 Aircraft K. Adère "Eole" [10].

However, because of the absence of lateral controls he failed to perform a full flight. But it was the first self-moving flight in history, carried out 13 years' prior to the Wright brothers' flight.

Aircraft conceptual development starts from the creative search stage: choice of style, composite approach, sketch creation, future model draw [13, 14]. In Fig. 1.14 the sketches of natural bionic forms and variants of creative search are shown. High-speed aircraft is assumed to be modeled and therefore as a final variant the construction with extended arrow-headed wing was chosen (Fig. 1.15, left side).

Further, the chosen sketch is modified based on aerodynamic operating characteristics and ergonomic requirements. After that the stage of 3D modeling based on detailed design takes place. In this work, graphic

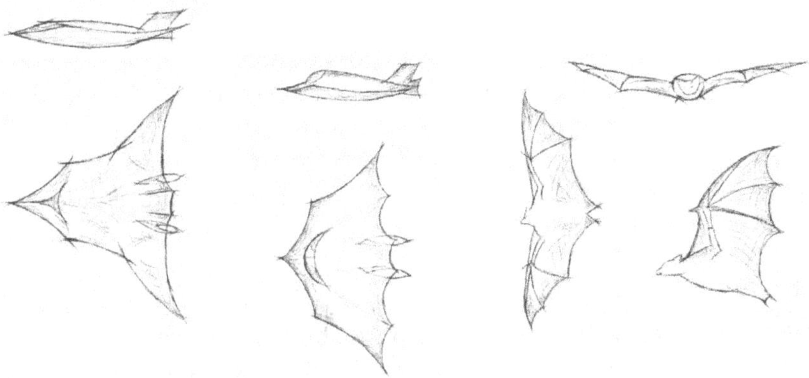

Fig. 1.14 Variant of creative search.

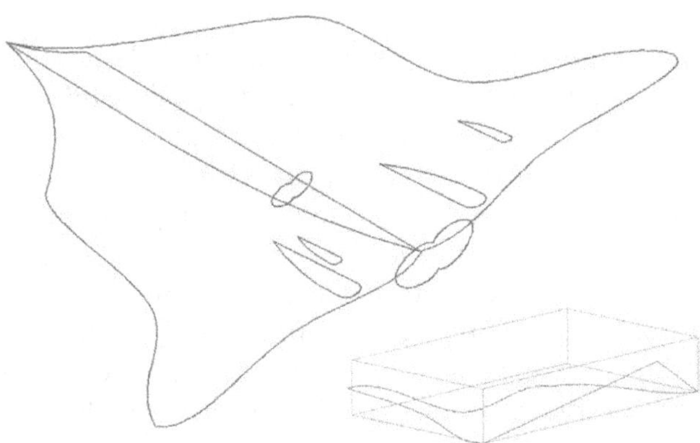

Fig. 1.15 Original plane forms of aircraft surface.

modeling system AutoCAD will be used. Nowadays there are a fair number of 3D modeling graphic systems and CAD system. Graphic system AutoCAD enables not only the creation of correct drawings, it can also carry out 3D surface, solid modeling and create realistic rendering. It should be noted that the author has considered the issues of amphibian aircraft 3D computer modeling in his papers [15, 38]. In the paper [38] the issues of amphibian aircraft conceptual design based on bionic forms were described.

To apply Koons method it is necessary to build overall 3D box [16]. 3D box is constructed by means of rectangle extrusion to some height. On each side edge of 3D box the curvelinear spline per closed loop is constructed (Fig. 1.15, right side, below). With the help of command Edge Mesh the constructed splines are pointed step by step, resulting in a smoothed curved surface.

While creating aircraft it is necessary to analyze the commands of surface modeling available as well as possibilities to design components of the object under modeling. We decompose aircraft (based on airframe axis): airframe upper and lower surface, vertical fin surface and engine surface. For better aerodynamic properties the aircraft airframe should be streamlined and for its creation the method of Edge Mesh will be used. Tail unit and engine nozzles can be created with command Ruled Mesh.

With the help of 3D box, the half of streamlined body along its axis is created. To create a triangle outline of the body upper part it is necessary to shift one of the closed spline peaks inside 3D box. Using Mirror tool, the wing splines mirrored copies are constructed. Original flatness for tail unit and nozzle is constructed in the form of polylines and a spline. The result of preparatory work to create aircraft surface in the form of flat splines and polylines is shown in Fig. 1.15.

Prior to performing 3D modeling let's consider design features of the aircraft model under development. Aircraft aerodynamic configuration pertains to "flying wing" type (without fuselage). As for arrangement, it is a monoplane; wing geometry in plan view is a variety of delta wing "ogival" type [17, 18]. Nowadays monoplane configuration with different wing arrangement is usually used. Delta wing is used mostly for supersonic aircraft. In comparison with straight wings they have a drag at cruise mode and are able to use wing inner volumes more effectively. Aircraft has no horizontal tail (tailless); vertical tail is a twin-finned one. Power plant consists of two turbojet engines located in vertical stabilizer section of the body. Landing gear configuration is supposed to be tricycle nose wheel landing gear.

In order to create aircraft 3D surface, the body's left and right parts are generated step by step with the help of Edge Mesh command. With the help of command Ruled Mesh the tail part and engines nozzles are created. The result of generation in three orthogonal and isometric projections is presented in Fig. 1.16. If necessary the body's original plane figures can be corrected for aircraft aerodynamic characteristics.

Further, in order to create accurate rendering scene it is necessary to assign the materials and determine lighting. The work with materials consists of two stages: adding (creating) new materials to drawing and material binding to scene objects. Creation of appropriate materials is an important part of shading process because the final result considerably depends on this. The material, in its turn, interrelates with lighting; object color affects specular reflection. In AutoCAD graphic system there is a big choice of materials; they can be edited or new ones created [16].

Notwithstanding that 3D models are more accurate than 2D ones they don't look natural; they lack real colors, shadows, lighting. Shading enables us to generate an accurate image of a 3D scene based on different optical effects. In Fig. 1.17 rendering scenes of aircraft "Chiroptera," shaded surfaces during cruising flight and target run are shown (Fig. 1.18). Design documentation can be also created in graphic system AutoCAD based on 3D model developed.

In conclusion it can be noted that in this paper the issues of new aircraft 3D modeling concept from creative sketches to accurate rendering were considered.

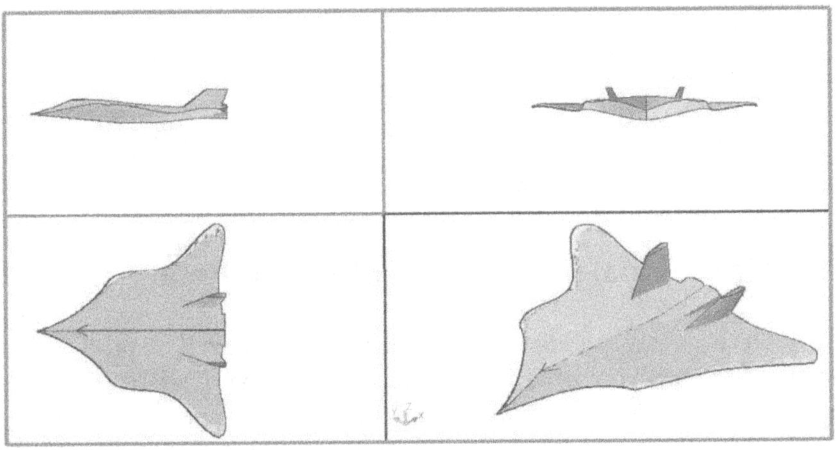

Fig. 1.16 3D surface model of aircraft in projections.

Fig. 1.17 Rendering of "Chiroptera" aircraft conceptual model (cruising model).

Fig. 1.18 Rendering of "Chiroptera" aircraft conceptual model (target run).

1.6 Conceptual Design of "Lotos" Motorcar

Today there are a lot of automotive giants in the world that produce new ranges of transportation with new technologies and possibilities each year. The companies embody the best solutions proposed by the designers. Design is the very beginning in the modern automotive industry. Each part of the car – from symbol to steering wheel – is carefully designed in the creative hands of the designers. It is body concept that gives birth to internal characteristics of the vehicle. The vehicle's outlook defines its "character".

Among the existing works on conceptual design we would like to note the articles [19, 20]. These works examine the importance of the sketching stage in the process of conceptual design of vehicles. Article [21] is dedicated to motorcar conceptual design; the stages of sketching, three-dimensional computer-aided modeling and final rendering of the scene are provided. The issues of initial stage of conceptual designing are described in detail in the book [22]. Provided there is the methodological basis of idea generation stage and determination of initial requirements for future structure.

This work is dedicated to three-dimensional computer-aided modeling of new vehicle concept. It should be noted that the issues of computer-aided modeling of airborne transportation means were reviewed by the authors in the works [15, 23]. The work [23] proposes conceptual visual-graphical solutions of new aircraft on the basis of bionic forms analysis.

The process of conceptual designing and modeling of transportation means takes several stages. At the first stage the sketch is created and a future model general view is drawn: composition solution; proportion of component parts relative to each other; main stylistic solutions. The concept of future prototype is chosen on the basis of analysis of natural forms visualization.

The concept of developed model design was determined by the lotus flower. Lotus (lat. *nelumbo*) refers to the kind of dicotyledonous plants, the only representative of the Nelumbonaceae family [24, 25]. Considering that the car design was inspired by this nice flower, the vehicle's outlook reveals the bionic form of lotus petals. In the culture of many ancient civilizations lotus signified immortality and the Divine, regeneration and the Sun, exhaustless energy and spiritual power.

When we talk about the automotive industry, people have different associations; for some people cars are quick, for other people they are compact, and for yet others they are spacious. But the first and basic thing is the company and its mark that is directly associated with the car. If we think about such trademarks as Lamborghini and Ferrari, we imagine their symbols at once in the form of a bull for the first and a horse for the second, and both animals are depicted in rather dynamic aspects. The success of many companies depends much upon the good quality design that predetermines the success due to recognition. Fig. 1.19 provides the

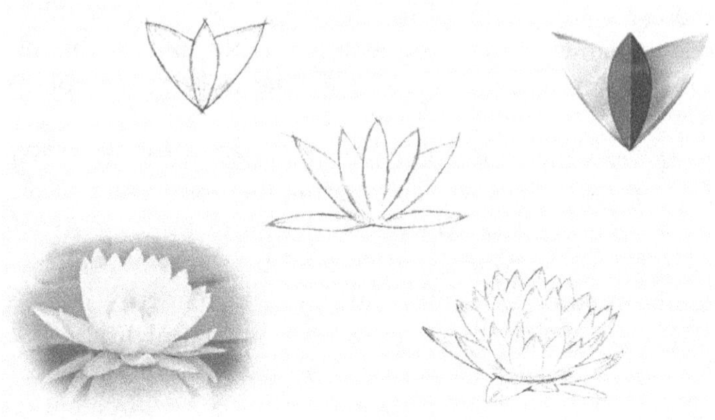

Fig. 1.19 Creative quest.

sketches of initial bionic forms and variants of creative quest of graphical symbol.

Many modern car makers have series with special characteristics. We are interested in series of fast high-performance cars that are powerful and graceful, such as: Maserati Quattroporte, Audi A7, BMW 5-series, Jaguar XF, Lexus LS [26]. These very models are close to "Lotos" model by ideology.

At the next stage the model is drawn with reference to man's biometric parameters considering the requirements of ergonomics. Fig. 1.20 shows

Fig. 1.20 Final variant of vehicle sketsh.

the final variant of future model sketch. Then the stage of modeling on the basis of the sketch project begins [27, 28]. In this work the graphical system of three-dimensional modeling 3ds Max will be used for modeling. 3ds Max graphical system is a flexible and multiplex software product that provides the user with a wide scope of operations [29].

Polygonal extrusion method is used for car model creation. Ortho graphic schematic drawings of conceptual model are located on projecting planes, as in Fig. 1.21. Object surface is created according to the car body projection.

Initially all component parts of the model are faceted (Fig. 1.21). The abilities of 3ds Max graphical system allow smoothing faceted objects by various methods. One of the variants is the application of smoothing method NURMS (Non-Uniform Rational Mesh Smooth). When car structural parts are constructed, the smoothing of polygons surfaces is performed (Fig. 1.22).

The next step of designing is shading and rendering of constructed three-dimensional model. The process of material assigning to separate structural parts of the vehicle is made on the level of polygons. When the operations are fulfilled, one can obtain a finished model for further rendering via realistic models of lighting. Integrated V-Ray module is used for

Fig. 1.21 Construction of body 3D model on the base of sketches.

Fig. 1.22 Three-dimensional smoothed surface of body.

Fig. 1.23 Rendering of "Lotos" conceptual model.

Fig. 1.24 Rendering of "Lotos" conceptual model.

rendering. Fig. 1.23 and Fig. 1.24 show final rendering scenes of shaded conceptual model of "Lotos" motorcar [30, 31].

Manufacturing of items and objects in our environment starts with the development of concept, prototype creation [32, 33]. In the result of modeling we can say that the concept of "Lotos" with the help of modern modeling systems was realized, starting from sketches to realistic rendering.

References

1. Abbasov, I.B., Lee, V.G., Ilyuschenko, N.L., *Three-dimensional modeling in the graphical system of the Mechanical Desktop* (tutorial), 112p, Publishing TRTU, Taganrog, 2004.
2. Abbasov, I.B., Computer simulation of the Be-200 amphibious aircraft. *Izv. SFedU.* Technical Science, 1, 160–164, 2009.
3. Hammond, D., Graphics in conceptual aircraft design - A designer's viewpoint, 1986. http://arc.aiaa.org/doi/abs/10.2514/6.1986-2733.
4. Haimes, R. and Drela, M., On the construction of aircraft conceptual geometry for high-fidelity analysis and design. *50th AIAA Aerospace Sciences Meeting Including the New Horizons Forum and Aerospace Exposition*, Article number AIAA 2012-0683, USA, January 2012, pp. 9–12, 2012.
5. McMasters, J.H. and Cummings, R.M., Airplane Design - Past, Present and Future. *J. Aircr.*, 39, 1, 10–17, 2002.
6. Liebeck, R.H., Design of the Blended Wing Body Subsonic Transport. *J. Aircr.*, 41, 1, 97–104, 2004.
7. Saeed, T.I., Graham, W.R., Hall, C.A., Boundary-layer suction system design for laminar-flying-wing aircraft. *J. Aircr.*, 48, 4, 1368–1379, 2011, doi: 10.2514/1.C031283.
8. Bolsunovsky, A.L., Sonin, O.V. *et al.*, Flying wing–problems and decision. *Aircraft Design*, 4, 4, 193–219, 2001. http://dx.doi.org/10.1016/S1369-8869(01)00005-2.
9. Gavel, H., Berry, P., Axelsson, A., Conceptual design of a new generation JAS 39 gripen aircraft. *Collection of Technical Papers - 44th AIAA Aerospace Sciences Meeting, Reno. USA, 9 - 12 January 2006*, vol. 1, pp. 395–406, 2006.
10. Website/Internet resource, Mode of access www/URL: http://en.wikipedia.org/wiki/Chiroptera (date access 12.08.2013).
11. Mosiyash, S.S., *A flying at night*, 160p, Znanie, Moscow, 1985.
12. www/URL: http://airwar.ru/enc/law1/eole.html (date access 12.08.2013).
13. Howe, D., *Aircraft Conceptual Design Synthesis*, 474p, Professional Engineering Pub. Ltd, London, 2000.
14. Jenkinson, L.R. and Marchman, J.F., *Aircraft design projects*, 371p, Butterworth-Heinemann, Oxford, 2003.

15. Abbasov, I.B. and Orekhov, V.V., *Amphibious. Computational modeling*, 69p, LAP Lambert Academic Publishing, Saarbrucken, Germany, 2012, (www.lap-publishing.com).
16. Abbasov, I.B., *Create drawings on the computer in AutoCAD 2012*, 136p, DMK Press, Moscow, 2011.
17. Yeger, S.M., Matvienko, A.M., Shatalov, I.A., *Basics of aircraft* Textbook, 720p, Mashinostroenie, M, 2003.
18. Raymer, D.P., *Living in the Future; The Education and Adventures of an Advanced Aircraft Designer*, 360p, Design Dimension Press, Los Angeles, 2009.
19. Eckert, C. and Stacey, M., Sources of inspiration: A language of design. *Design Studies*, 21, 5, 523–538, 2000.
20. Bouchard, C., Aoussat, A., Duchamp, R., Role of sketching in conceptual design of car styling. *J. Des. Res.*, 5, 1, 116–148, 2006, doi: 10.1504/JDR.2006.010810.
21. Damujanovic, D., Kozak, D., Ivandic, Z., Kokanovic, M., Car design as a new conceptual solution and CFD-analysis in purpose of improving aerodynamics. *Fisita World Automotive Congress, Budapest, Hungary, 30 May – 4 June 2010*, Vol. 1 of 5, pp. 3877–3885, 2010.
22. Kroll, E., Condoor, S.S., Jansson, D.G., *Innovative Conceptual Design: Theory and application of parameter analysis*, 227 p, Cambridge University Press, 2001.
23. Abbasov, I.B., Conceptual model of aircraft "Chiroptera". *Am. J. Mech. Eng.*, 2, 2, 47–49, 2014, doi: 10.12691/ajme-2-2-3.
24. Martin, W., *Wild Flowers*, p. 512, Harpercollins Pub Ltd, 1999.
25. Website/Internet resource, Mode of access www/URL: http://wikipedia.org (date access 19.03.2015).
26. Tarasov, R., Review of executive cars. Car newspaper, 6, Klaxon, 2010, http://www.klaxon.ru (date access 17.03.2015).
27. Vasin, S.A., Talaschuk, A.U. et al., *Design and modeling of industrial products*, 692p, Mashinostroenie, Moscow, 2004.
28. Abbasov, I.B., *Computational modeling in industrial design*, 92p, DMK Press, Moscow, 2013.
29. Abbasov, I.B., *Basics of three-dimensional modeling in the graphics system 3 ds Max 2009*, 176p, DMK Press, Textbook. Moscow, 2010.
30. Gabrilyan, G.V., Graphic accompaniment of the automobile series "Lotos". Graduation project under the guidance of Abbasov I.B, 57p, TIT SFU, Taganrog, 2012.
31. Abbasov, I.B. and Gabrilyan, H., Conceptual Design of 'Lotos' Motorcar. IOSR *J. Comput. Eng.*, 18, 2, Ver.4, 33–36, 2016. doi:10.9790/0661-1802043336, http://www.iosrjournals.org/iosr-jce/pages/18(2)Version-4.html.
32. Runge, V.F. and Manusevich, Y.P., *Ergonomics in environmental design*, 328p, Architecture-C, Moscow, 2005.

33. Happian-Smith, J., *An Introduction to Modern Vehicle Design*, 600p, Elsevier Limited, 2002.
34. Abbasov, I.B., Computer simulation of the aircraft. Proceedings of the V International STC "Computer modeling". Part 1. S.Pb. SPbGPU, pp. 207–209. 2004.
35. Abbasov, I.B., *We create drawings on a computer in AutoCAD 2007/2008* (Textbook) (second edition), 136p. M.: DMK Press, 2007.
36. Abbasov, I.B., *We create drawings on a computer in AutoCAD 2012* (Textbook) (third edition), 136p. M.: DMK Press, 2011.
37. Abbasov, I.B., Computer simulation of streamlined surfaces. Proceedings of the IV International STC "Computer modeling". St. Petersburg. SPb GPU. pp. 255–256, 2003.
38. Abbasov, I.B., *Drawings in a graphical environment AutoCAD: Tutorial*, 82p, Publishing house TSURE, Taganrog, 2002.

2

Conceptual Modeling of Amphibian Aircrafts

Iftikhar B. Abbasov* and V'iacheslav V. Orekhov

Southern Federal University, Engineering Technological Academy, Department of Engineering Graphics and Computer Design, Taganrog, Russia

Abstract

The second chapter presents 3D computational modeling of amphibious aircrafts Be-200, Be-103. Hereby the process of amphibious aircraft components phased modeling is presented. Examines issues regarding computational modeling of multipurpose passenger amphibian aircrafts Be-200, "Lapwing" cabin interior. Here different concepts of cabin layout are introduced: economy variant; comfortable layout; with coupe-type seating; corporate variant with berths. Objects interior is designed on the basis of ergonomic principles. For cabin computational modeling the 3ds Max graphic system is used. Objects modeling is carried out by means of Spline Extrude, Polygon Extrude methods. In course of scenes shading, the materials assignment is performed at the level of sub-objects. Scenes of realistic rendering of various aircraft cabin layouts are introduced.

This is dedicated to computational modeling of the conceptually new amphibious aircraft "Lapwing". Based on the analysis of bionical forms of operational medium, the visual and graphical solutions of the developed model are provided. Sketch drawings considering the requirements of ergonomics are provided; a sketch of amphibious aircraft 3D model is created. Based on sketch projects the stage-by-stage 3D modeling of amphibious aircraft structural parts was performed.

This chapter is devoted to computer modeling of conceptually new ekranoplan "Water Strider" and design of its passenger salon, from the sketch to ready model. In the works features of this type of transport are noted and a review of the current state of the market of small-size ekranoplan of different function is made. Publications concerning a research and model operation of the vessels using

**Corresponding author*: iftikhar_abbasov@mail.ru

Iftikhar B. Abbasov (ed.) Computer Modeling in the Aerospace Industry, (23–136)

wing-in-ground effect during flight are described. The concept of a new ekranoplan on the basis of a biological prototype from the operation environment is offered; visual and graphic solutions of the developed model are provided. Outline drawings, the key operational parameters of an ekranoplan are presented, and a stage-by-stage three-dimensional model of design parts of the vessel is carried out. Explicitly, stages of modeling of a passenger seat, creation and assignment of materials when shading a scene are considered, characteristics of sources of lighting, feature of rendering process are specified. The final scenes of realistic rendering are presented: rise of an ekranoplan, an interior of passenger salon, the coastal landing module for an ekranoplan.

A prototype hydrofoil "Afalina" was developed; the main features of this type of vessel and their multifunctionality are described; the design process is presented starting from the concept to the final rendering of the finished model.

This chapter is dedicated to computer modeling of the autonomous mobile robotic system "Sesarma". The review of publications and analogues is performed; the structure diagram and design of the developed robotic system are presented. The search for robot body concept is carried out, the process of setting and assignment of materials based on polygonal model elements is described. Renderer is selected and light sources are installed, and final scenes of scene rendering with robot are shown.

Keywords: Computer modeling, amphibious aircraft, ekranoplan, hydrofoil, bionics, mobile robot, ergonomics, cabin interior

2.1 From the History of World Civil Aviation

2.1.1 Introduction

Civil aviation has been developing at a high pace. Aircraft, unlike the other means of transportation, allow their passengers to save time when covering any distance. In Russia, with its vast distances between the business and cultural centers, civil aviation plays an exclusive role. Its condition has significant influence on economic efficiency.

It should be noted that the rivers and lakes in Russia have traditionally been of special importance, not only as transportation routes, but also as the routes of public settlement and development of new territories. Nearly all the big cities are built on river banks. With this amount of water areas available, it is most reasonable to use them as a transportation network.

Little by little the development of civil aviation reduced the necessity of passenger transportation by water. And yet we still have inhabited areas reachable solely by waterways. This is where hydroaviation comes to the aid of the people. As for the hydroplane, its application is limited by the water area, while the amphibious aircraft can be operated both from the water area and the earth.

It was historically predetermined that the hydroaviation development and application concept in Russia and in most developed countries would diverge from mainstream aviation. Industrially developed regions with their vast airfield infrastructure do not require the application of water-based flying vehicles or amphibians. They are more expensive and complicated in terms of operation and maintenance. In many countries, hydroaviation is mostly used as an exotic means of transportation for the purpose of tourism and recreation. Due to this, hydroplanes, amphibious aircraft and wing-in-ground effect vehicles constitute an insignificant part of the world aviation fleet.

As of today, many aviation companies deal with the development of hydroplanes. Major among them are Canada's Canadair, Japan's Shin Meiwa, China's Harbin Hafei Airbus and the Beriev Aircraft Company, Russia. The Beriev Aircraft Company has a 70-year experience in hydroplane development and deservedly occupies the leading position in world hydroaviation.

This paper briefly describes the history of hydroplane development by the Beriev Aircraft Company. Further, it reviews the issues of 3D computer modeling of amphibious aircraft Be-200 and Be-103, introducing the main stages of 3D computer modeling. The paper reviews their structural peculiarities and describes the amphibious aircraft major parts modeling process using different methods. Shading and material assignment are done at the sub-object level. The paper represents also the realistic rendering of the amphibious aircraft 3D models.

2.1.2 Historical Stages of Hydroaviation Development by the Beriev Aircraft Company

In 1920 the Soviet state, despite the civil war, created seadromes and sea aviation schools in Peterhof, Nizhni Novgorod, Samara, Nikolaev and Odessa. The following home-made aircraft were created: MR-1, ROM-1, MR-5, POM-1, TB-1a, MDR-3, MBR-2, KOR-1. The basic seaplanes MMR-317, MDR-301, KORP-3, Che-2, MTB-2, and others made an important contribution during the Great Patriotic War [1].

Soviet hydroaviation was greatly influenced by the famous aircraft designer A.N. Tupolev. Along with the development of land-based aircraft, he conducted activities on the creation of hydroplanes: ANT-8; ANT-22, ANT-27. The last hydroplane designed by Tupolev Design Bureau was the sea flying boat ANT-44 (tested in 1937).

In 1934 Central Design Bureau of Seaplanes Manufacturing headed by Georgy M. Beriev as the Chief Designer was set up in Taganrog at the aviation plant facilities [2]. The Design Bureau started its activity with the serial production of hydroplane MBR-2. Both cargo and passenger versions were constructed: MP-1 and MP-1T. These aircraft became the basic representatives of sea aviation in Russia.

After the liberation of Taganrog from the German troops, hydroplane construction was resumed. In 1944 hydroplane LL-143 was put into production, in 1946 long-range hydroplane Be-6 (Fig. 2.1.1) was developed [3]. In 1948 light amphibian Be-8 was developed, constructed and underwent flight testing (Fig. 2.1.2).

The development of jet aviation raised the issue of constructing an aircraft with turbo-jet engines, and such an aircraft was developed in 1952. Jet flying boat R-1 (Fig. 2.1.3) got off the ground for the first time that year [4].

Fig. 2.1.1 Hydroplane Be-6.

Fig. 2.1.2 Light amphibian Be-8.

Fig. 2.1.3 Jet flying boat R-1.

In 1961 an aircraft with turbo-jet engines, Be-10, was represented (Fig. 2.1.4). This hydroplane made a sensation in the world. It set up 12 world records including the speed record of 912 km/h.

In 1960 hydroplane Be-12 (Fig. 2.1.5) passed the tests, and was put into serial production later, in 1963 (the Chief Designer was A.K. Konstantinov). The amphibious aircraft A-40 "Albatross" (Fig. 2.1.6) made its maiden flight in 1986. This is the world's largest amphibious aircraft distinguished by unique flight performance and seaworthiness characteristics; it set up 144 world records.

In 1989 Beriev Design Bureau started working (the Chief Designer was G.S. Panatov) on the creation of multipurpose amphibious aircraft Be-200

Fig. 2.1.4 Flying boat Be-10.

Fig. 2.1.5 Flying boat Be-12.

Fig. 2.1.6 Amphibious aircraft A-40 "Albatross".

(Fig. 2.1.7). The experience of A-40 hydroplane development was widely used in its designing. Be-200 prototype made its first flight in 1998 [5]. In June 1999, it was exhibited at Le Bourget air show [2].

Externally Be-200 resembles the scaled-down A-40 (the length and the wing span are 25% less, the takeoff mass is twice smaller). The amphibian operation is not limited to the high-class airfields, which makes it irreplaceable in the Far North, Siberia and the Far East regions. The basic functionality of Be-200 is forest fire-fighting.

Fig. 2.1.7 Amphibious aircraft Be-200 takeoff from shallow water surface.

In 1998, after passing the tests, the light amphibious aircraft Be-103 (Fig. 2.1.8), intended for the transportation of 5-6 persons, was represented. The aircraft is distinguished by an uncommon, for this class of aircraft, layout: monoplane with a low-mounted water-displacing and hydroplaning wing. The aircraft is designed for short air routes in coastal and island states.

To date, a number of modifications of this new amphibian have been developed in addition to the basic passenger version. Be-103 can be used for passengers, mail and small-cargo transportation, fire-fighting surveillance, patrol and ecologic monitoring of water areas, rendering first medical aid, and search-and-rescue operations.

Fig. 2.1.8 Amphibious aircraft Be-103.

2.2 Computational Modeling of Multipurpose Amphibious Aircraft Be-200

2.2.1 Introduction

Day by day the necessity for visual images of projected industrial products in different spheres of human activity increases. From the economical point of view it is justified. State-of-the-art technologies of computational modeling and design enable the achievement of results which engineers could not even dream of just 20 years ago [6].

Application of modeling state-of-the-art technologies for aircraft designing is of interest at this time. Let's scrutinize certain current articles dedicated to the computer modeling of airplanes. In the paper [7] the issue regarding high-precision software tools implementation for design process optimization is considered. Capabilities of new software when aircraft structure designing are described.

The paper [8] is dedicated to aerodynamic configuration of hypersonic transport conceptual design based on multipurpose optimization of structure. In papers [9, 10] the issues regarding cost-effective subsonic passenger aircraft development are considered. The paper [11] is dedicated to conceptual design of aircraft where bird wings aerodynamic quality is taken into account.

In the paper [12] the development of wing geometry design based on its flexibility and effective aerodynamic load is carried out. Here, the effect of different materials on the wing structure stiffness is described. The issues of airplane designing based on the conic lofting-based software are investigated in the article [13], which performed the design of the wing geometry and examined its aerodynamic properties.

Software for parameter designing of the airplane aerodynamic surfaces is introduced in the article [14]. The streamlined and curved structural members are generated on the basis of surfaces NURBS. The feasibilities of special software for designing of different airplane configuration are set forth in article [15]. The proposed configuration design tool could be especially efficient when automation, flexibility and rapid changes of geometry are required in a short time and with low computational resources.

It is to be noted that in the past decades the necessity for emergency-rescue aviation appeared. In conditions of natural disaster, technogenic catastrophes it is impossible to manage without aviation emergency support. Along with land-based aviation, amphibious aircraft are used with increasing frequency. As an example we can provide fire sea planes [1], which are essential during the extinguishing of huge forest areas.

This paper is dedicated to computer-based 3D modeling of amphibious aircraft Be-200 developed by Beriev Aircraft Company (Fig. 2.1.7). The Be-200 aircraft design process was started in 1990 on the basis of amphibious aircraft A-40 "Albatross". Be-200 differs from it with less size. the project was introduced for the first time in 1991 during the Paris aerospace exhibition.

Amphibious aircraft Be-200 configurations are as follows: firefighting, passenger and cargo transportation. A fire aircraft is equipped with water tanks with total capacity of 12 m^3, the filling of which is carried out in gliding mode over water surface by means of water scoops located behind the boat step [2].

The tanks' lower portions are provided with automatically opening doors to drop water. With takeoff weight 37 200 kg the aircraft is capable of carrying up to 12 000 kg of water to a fire located at a distance of 100 km from the airdrome and 10 km from the basin. Amphibious aircraft Be-200 wing span is 32.78 m, length 32.05 m, height 8.9 m, fuselage maximum diameter 2.86 m, air crew 2 persons. Maximum cruise speed at altitude 8000 m is 710 km/h, distance of takeoff from water is 1000 m.

2.2.2 Modeling Methods and Stages

Currently there are a fair number of 3D modeling graphic systems and CAD system. For amphibious aircraft modeling we will use 3D modeling graphic system 3ds Max. Along with many 3D modeling graphic systems 3ds Max practically enables to develop designs of any complexity. Graphic system 3ds Max is a versatile software product. It cooperates with many engineering applications providing a wide work scope for the designer [16].

In order to start the work, initial data like drawings, figures, layout projections of the object under modeling are required (Fig. 2.2.1). Further, three mutually perpendicular planes are created in graphic system 3ds Max with projection images on them (Fig. 2.2.2) [17].

To develop 3D model there are some methods, one of which is a polygonal extrude. In 3ds Max graphic system polygonal modeling has a number of special features. For more precise modeling of fuselage (boat) lines we mentally split fuselage along fore-and-aft axis. Sections received are placed in scene according to drawing (Fig. 2.2.3).

Graphics-oriented system 3ds Max contains various types of editable objects; among these is Editable Poly. This object includes the following sub-objects: vertex; segment; border; polygon; element. Polygon sub-object level is chosen for application of Extrude method. The content of the

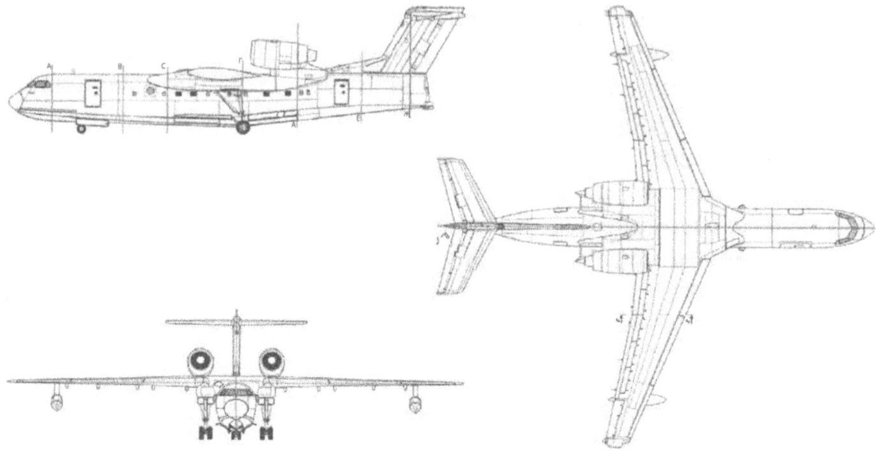

Fig. 2.2.1 Orthogonal projections of amphibious aircraft Be-200.

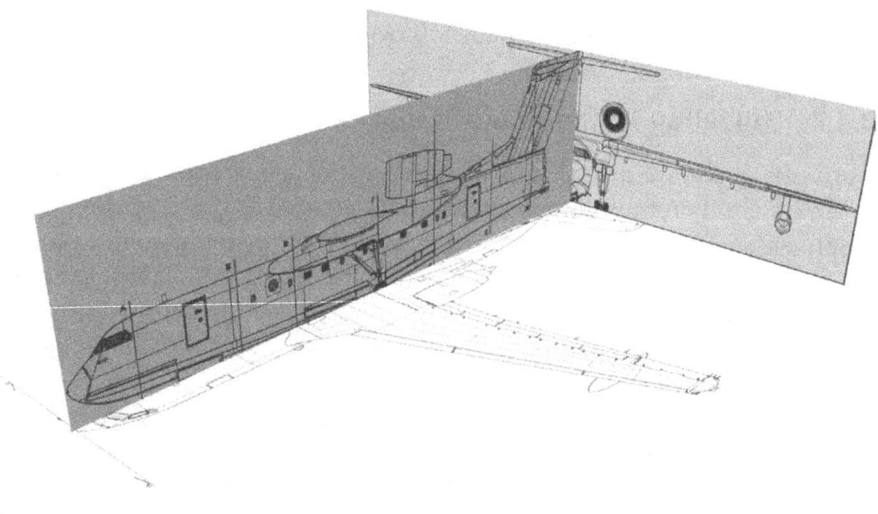

Fig. 2.2.2 Construction of mutually perpendicular planes.

method is in a consequential extruding of the selected polygons longwise of their own standard part. Besides, to make work easier, one can be carried out with one half along fore-and-aft axis of the future object (in our case it is the right half, and the left one will be automatically built at the final stage) [18].

The number of polygons should remain constant over the whole model; thus the work starts from cylindrical work with a definite number of

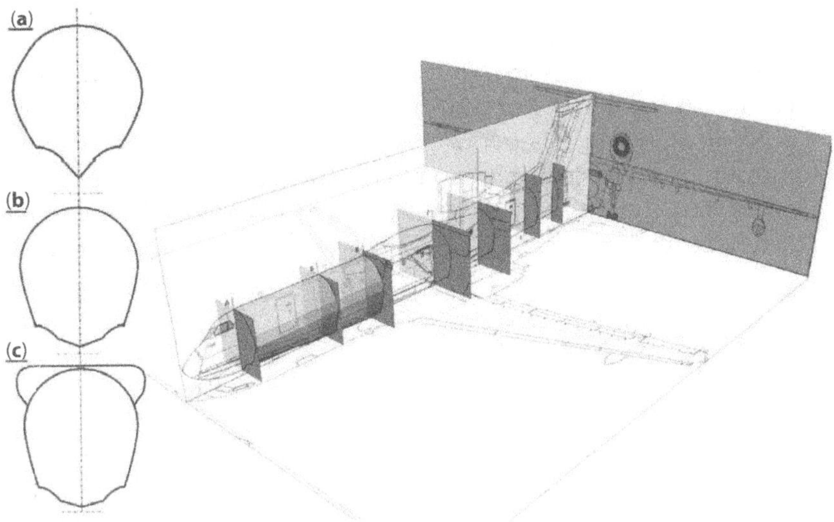

Fig. 2.2.3 Supporting sections arrangement of fuselage.

segments. Fuselage extruding on the basis of supporting sections is shown in Fig. 2.2.4. Upon application of the command Extrude the obtained increment of basic object should edit manually taking into account the orthography contours. Hand finishing is to be carried out at the level of vertex sub-object, because this is a time-consuming process and requires

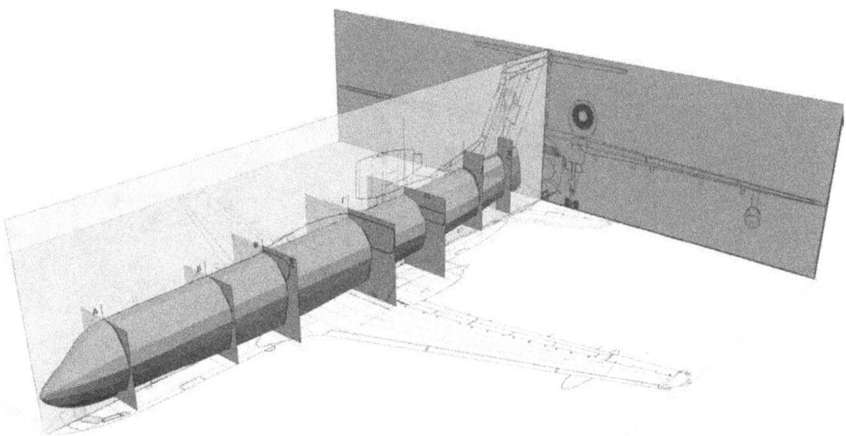

Fig. 2.2.4 Sequential polygonal extrude of fuselage.

constant coordination of all planes simultaneously. Gradual wing surface extrusion is shown in Fig. 2.2.5.

All model components are created similarly: engine body, tail unit, stabilizer's float (Fig. 2.2.6). Further, the mirror image reproduction of the constructed model half is performed with use of command Symmetry taken from a list of accessible modifiers. The detailed final working out of airplane structure is carried out at the final stage [19, 20], the result of simulation is given in Fig. 2.2.7.

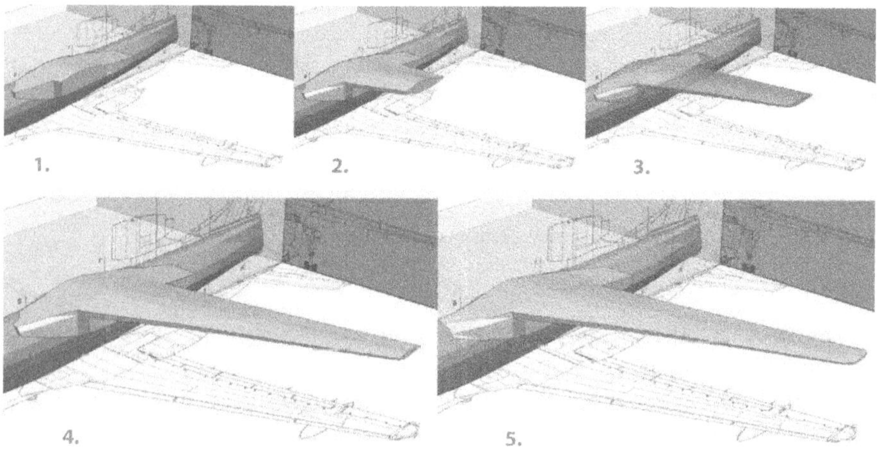

Fig. 2.2.5 Gradual wing surface extrusion.

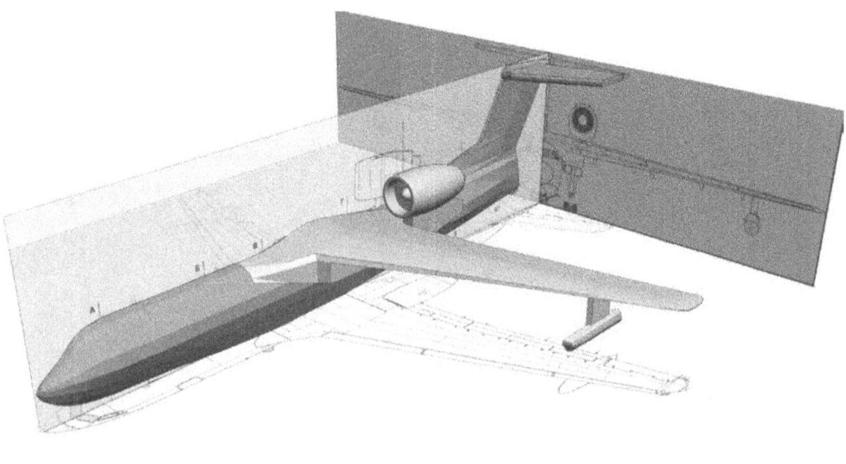

Fig. 2.2.6 Sequential polygonal extrude of engine and stabilizers shaping.

Fig. 2.2.7 3D model of amphibious aircraft Be-200 assembly.

2.2.3 Shading of 3D Model

Objects surrounding us have essential properties, which identify their appearance. To get the actual photographic images in graphics-oriented system 3ds Max the choice of material and customization of its main properties are to be the defining ones. Hereafter, for the sake of rendering, the V-RAY rendering unit will be used. It's stipulated with use of materials VRayMtl type.

Window of Editor Materials VRayMtl differs from the standard one. Let's scrutinize the possibilities of this edit program in terms of creation of the fuselage surface material. One of the significant material parameters is its main color (diffusive color); here the body color is white. Under the icon Diffuse there is a window Reflect, which is responsible for the reflection properties of the material chosen. Black means that material is nonreflective, whereas white means the full reflection. In this case to create the effect of the painted glossy finish it's required that the degree of surface reflection would depend on the review angle (Fresnel effect). The Falloff procedural is used to receive the same (Fig. 2.2.8).

Reflection glossiness parameter affects the material accurateness as well. During change of reflection glossiness parameter value from 1.0 to 0.9 and reflections become fuzzy (Fig. 2.2.9). The same effect permits the creation on the surface of very small noise of finish in the form of inequality, scratches.

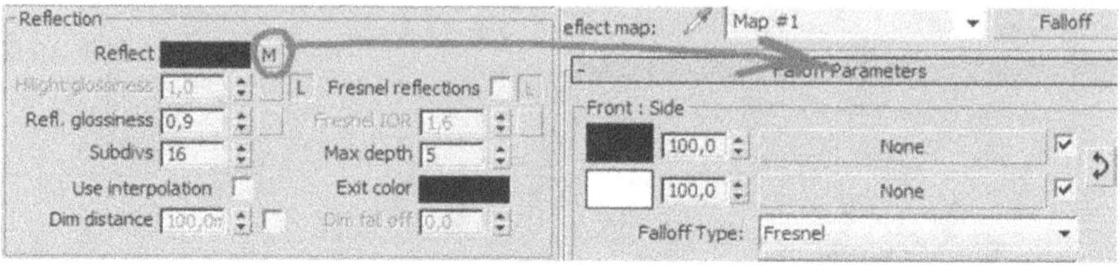

Fig. 2.2.8 Use of falloff procedural card for reflect parameter.

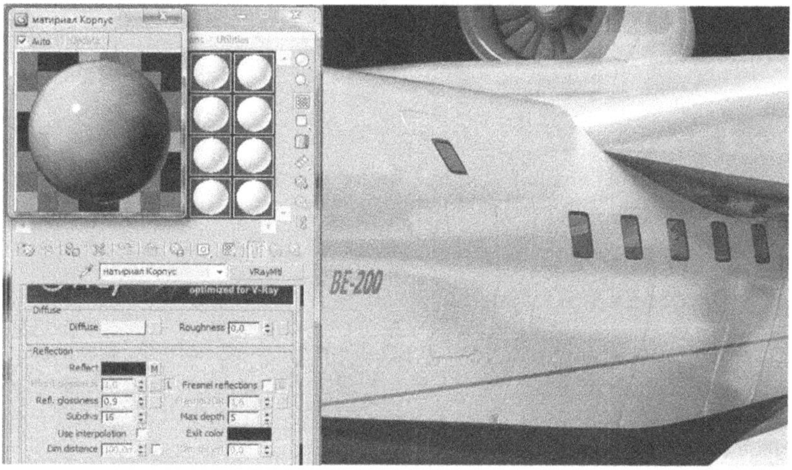

Fig. 2.2.9 Fuselage material in the window Editor Materials.

2.2.4 Rendering of 3D Model

The important detail of any scene is lighting. In general, in almost all scenes one of two types of lighting is used – natural and artificial. The best way to create natural lighting is to use lights radiating one-way parallel rays of light. This type of lighting can be created through directional light. Generally, artificial lighting is created by means of few lights with low intensity.

Scene objects can be rendered with different degrees of precision. In graphic system 3ds Max several layout engines are used: for objects view in viewport, for materials sketches view and for finite picture getting. These engines enable to find trading time for quality. Graphic system 3ds Max has a module for rendering but the capabilities of standard visualizer do not provide a real picture. Module V-RAY was used for rendering of this work. Its high possibilities enable the receiving of a more precision picture.

(a)

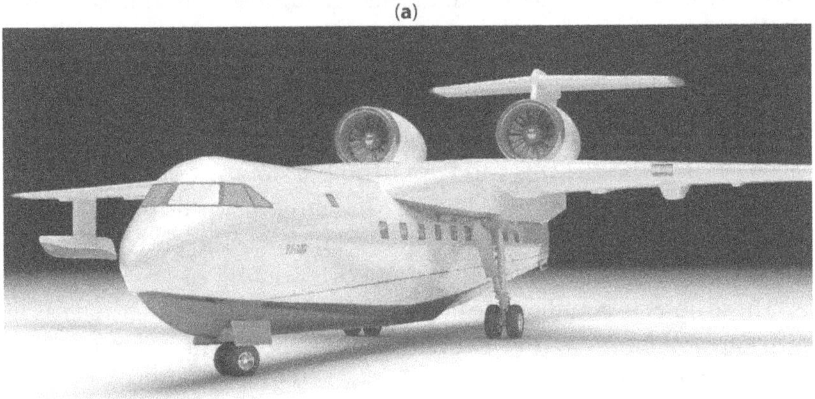

(b)

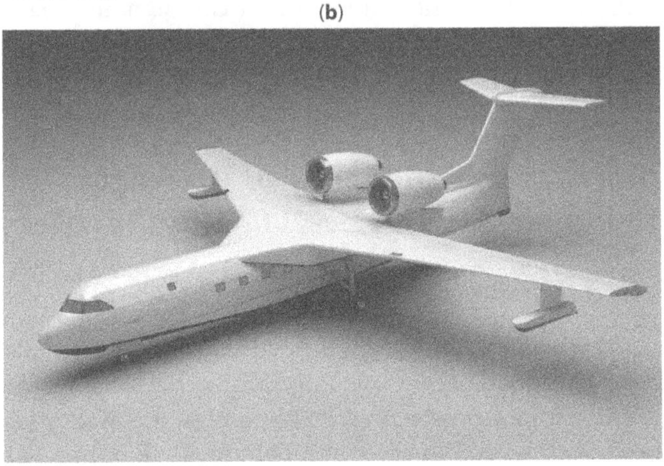

(c)

Fig. 2.2.10 Rendering of shaded amphibious aircraft Be-200 model.

The principle of V-RAY work is in indirect illumination; it's related to the algorithm of modern rendering modules. Indirect illumination is an object's lighting only by diffusely reflected light from other objects without a direct one from the direct light source. For receiving of further actual photographic images, the scene along with airplane model was created and set in a certain manner. On renderings the illusion is created that the airplane is in the dark hangar with spotlight lighting on the ceiling (Fig. 2.2.10, (*a*), (*b*), (*c*)). This result was received due to the installation of several light sources V-RAY Light above the model. Daylight is missing in the scene, but the indirect illumination algorithm permits the avoidance of the dark areas of the image without usage of additional illumination of the model underneath.

2.2.5 Conclusion

It will be noted that in this modeling the possibilities of the current 3D simulation graphic system were used. Modeling methods used in the creation of 3D objects are to be the well-known ones. The described method is not considered to be an exclusive one for solving of the design problems. These objects could be created with the use of other methods as well. Choice of modeling method depends on the possibility of software and designer's preference.

As a result of developments carried out it can be noted that rendering scenes of amphibious aircraft Be-200 3D model are considerably photorealistic and provide rendering of the object designed.

2.3 Computational Modeling of Passenger Amphibian Aircraft Be-200 Cabin Interior

2.3.1 Introduction

In countries with continuous waters, hydroaviation can make a good contribution. In modern economic conditions the development of civil hydroaviation becomes prospective. In spite of world wars, the development of the civil aircraft industry continued throughout the 20th century. During this time aircraft design was improving, the necessity of increasing passenger capacity was growing, and ferry time was declining. With relation to fuselage design, the main trends of shape forming were developed already by the 1960s. Since that time fuselage shape and contours have remained the same in general. Therefore at the present time works regarding fuel consumption, cabin noise, and aircraft structures weight reduction are carried out.

In Russia the main developer of hydroaviation is a Beriev Aircraft Company, Taganrog. In 1990, the Beriev Aircraft Company started the amphibian aircraft Be-200 designing (Fig. 2.3.1) [2]. Initially the project was introduced in 1991 during Aeroshow in Paris.

Features of amphibian aircraft Be-200 3D computational modeling were examined in papers [18, 21]. In the paper [22] the process of the said aircraft phase modeling is described, and variants of developed model rendering introduced (Fig. 2.3.2). This paper is concerned with passenger amphibian aircraft Be-200 cabin layout and computational modeling. To date, Be-200 passenger version offered by the manufacturer is at the design stage [2]. The authors of this paper propose new original concepts of the passenger cabin layout intended to satisfy different demands.

Aircraft cabin design issues remain important for civil aviation. Some aspects of cabin design and passengers comfort during flight are described in papers [23, 24]. In paper [25] the future passenger aircraft cabin design concepts are reviewed. Different issues of aircraft cabin optimal layout on the basis of current design specifications are described.

A rather detailed description of these issues in the paper [26] should be noted. This book is concerned with fuselage configurations, their advantages and disadvantages. Practical recommendations regarding cabin designing based on fuselage structure and ergonomic norms are introduced.

Fig. 2.3.1 Amphibian aircraft Be-200 takeoff from water.

Fig. 2.3.2 Amphibian aircraft Be-200 computational model rendering.

2.3.2 Variants of Cabin Layout

Inner space designing of any type of transport starts with tasks definition: purpose of development and operation conditions. Thereon hangs the general architecture of the future project. When designing passenger transport, there are a number of key points: organization of passenger seats, lavatory, galley and cargo compartment availability and configuration [27]. In this case, volume and overall dimension is limited but sufficient for comfort use. Ergonomic requirements to passenger chairs, lavatories and other objects should be taken into account.

Amphibian aircraft Be-200 has the following versions: firefighting, passenger, cargo [2]. Wing span of amphibian aircraft Be-200 is 32.78 m, length 32.05 m, height 8.9 m, fuselage maximum diameter 2.86 m, crew 2 persons.

In our case the task is to convert the current structure of amphibian aircraft Be-200 into a variant of a comfortable passenger cabin. For this purpose it is necessary to organize space, meeting all necessary requirements, both aesthetical and practical. Materials hygienic properties used in decoration also should be taken into account. In-flight injury prevention depends on correct ergonomics.

We will start aircraft cabin designing from the development of layout diagram which is defined upon the task assigned. The suggested commercial variant of the passenger aircraft has a capacity for 72 passengers and 4 crew members [28]. This variant of cabin provides maximum number of passenger seats without loss of comfort for any particular passenger (Fig. 2.3.3). It should be noted that this layout of the passenger cabin

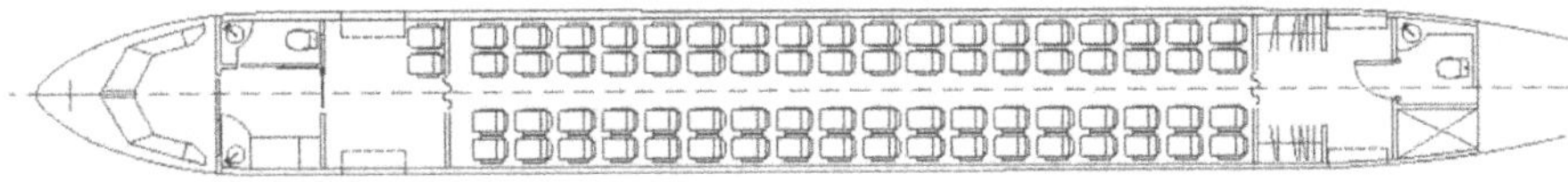

Fig. 2.3.3 Layout scheme of aircraft cabin economical variant.

was developed by the manufacturer, and yet this design has not been implemented.

For this layout diagram, there are several possible variants of passenger chairs organization in cabin: standard and maximum commodious. In this case, chairs are forward-facing, in two rows with an aisle between them (Fig. 2.3.4).

For coupe layout diagram the chairs are located facing each other with folding table possibility (Fig. 2.3.5). Number of passengers – 56, crew and servitorial staff – 4 (Fig. 2.3.6).

Making comparison with railway transport, the variant of cabin commercial layout of enhanced comfort can be noted. It provides passengers accommodation in individual berths with minimum set of equipment required: sleeping place, compartments for passenger cabin baggage, multimedia equipment (TV, Wi-Fi).

A more comfortable layout version that accommodates fewer passengers but provides higher comfort is proposed for review below. This

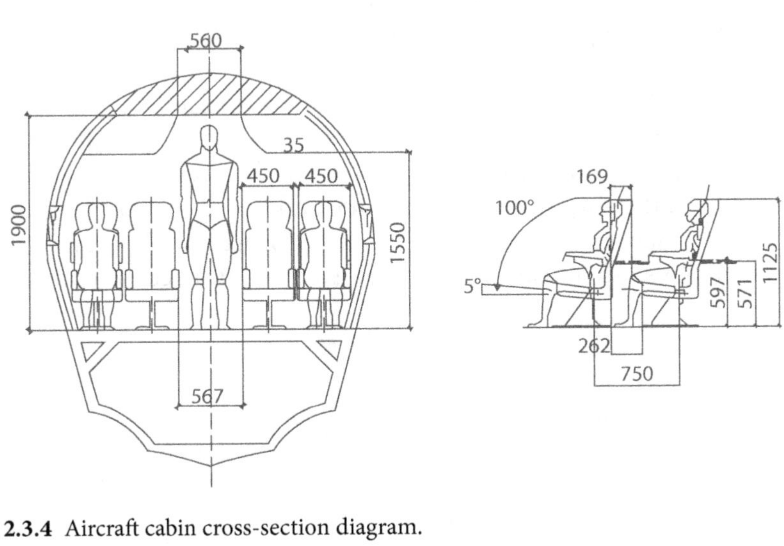

Fig. 2.3.4 Aircraft cabin cross-section diagram.

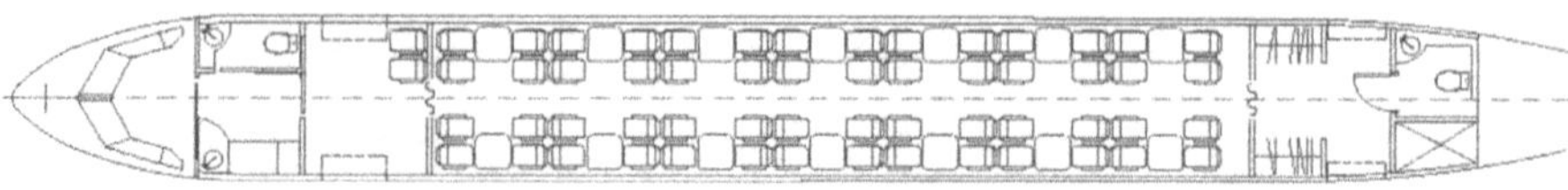

Fig. 2.3.5 Layout scheme of chairs with coupe-type location.

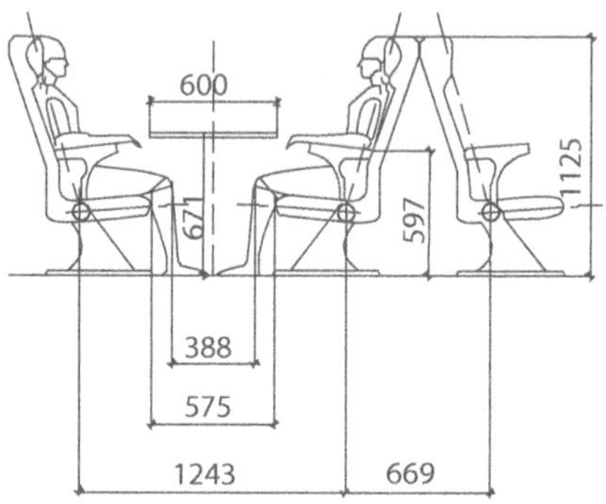

Fig. 2.3.6 Chairs layout in cabin, coupe-type variant.

layout stipulates the location of seats in one row along the aircraft sides (Fig. 2.3.7).

This layout solution allows the accommodation of three lavatory units, one for the crew, and two for the passengers, mini galley-unit, and a bar. The number of passengers to be accommodated amounts to 22, plus three crew members, including a flight attendant. Fig. 2.3.8 shows the fuselage cross-section and a passenger seat configuration to give an idea about the cabin dimensions.

Enlargement of the seat dimensions allows location of two rows of seats with a comfortable aisle between them. Together with the enlargement of transverse distance, it is proposed to increase the longitudinal pitch between the seats in order to provide additional comfort (Fig. 2.3.8). The distance between the seats enables transformation of every seat into a comfortable sleeper seat. Such cabin layout allows accommodation of

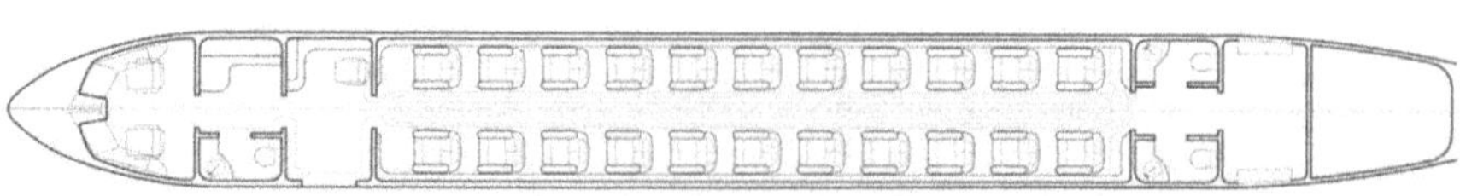

Fig. 2.3.7 Layout scheme of aircraft cabin comfortable variant.

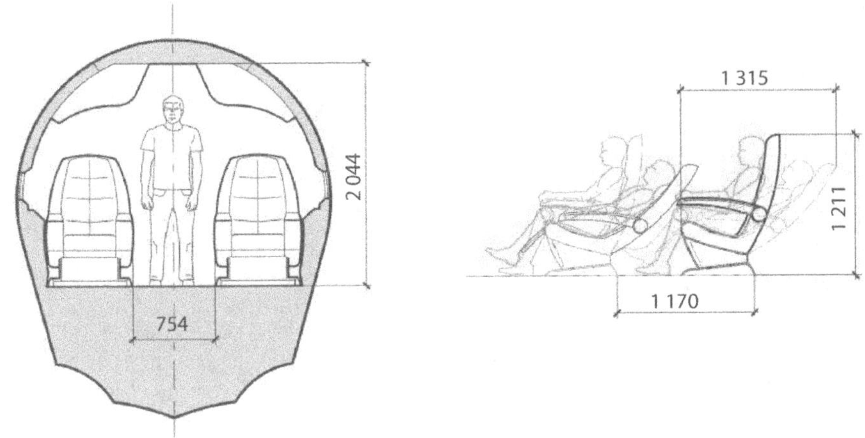

Fig. 2.3.8 Fuselage cross section and seat ergonomic parameters.

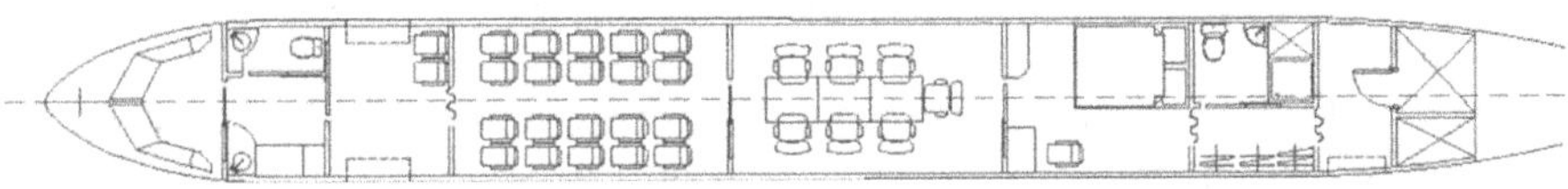

Fig. 2.3.9 Layout scheme of VIP-cabin.

individual ventilation and temperature control panels and of a multimedia unit for headset connection.

Besides passenger variant the corporate variant with VIP-cabin providing the availability of separate apartments is possible. This layout includes a separate lavatory along with shower cabin, comfortable and wide bed, and a room for meetings as well. This variant can be manufactured both for businessmen and for top public officials (Fig. 2.3.9).

2.3.3 Aircraft Cabin Modeling

For fuselage and cabin realistic model creation, we will use layout diagrams, section views, and cross-sectional views of the future hull developed earlier. 3D computational modeling of the cabin will be created with the help of 3ds Max graphic system. 3ds Max graphic system is a universal tool of modeling and alongside many other 3D systems enables the development of projects of any complexity [16]. In our case we will create fuselage lines on the basis of splines extrusion method Spline Extrude (Fig. 2.3.10).

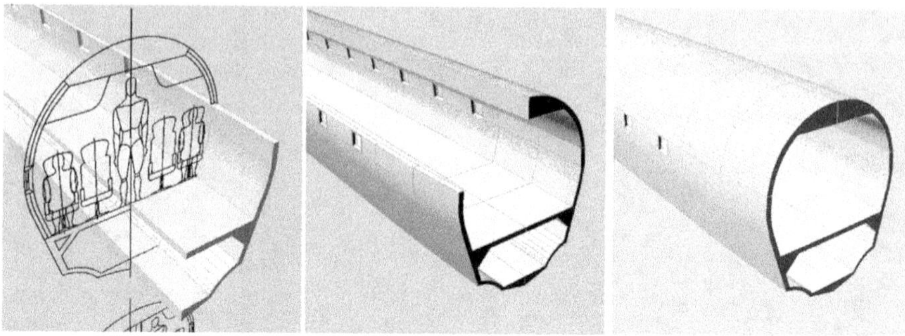

Fig. 2.3.10 Successive steps of cabin modeling.

For detailed elaboration, the cabin main elements are to be created separately from the whole scene. Modeled elements are arranged over cabin by means of duplicating. All cabin elements are created the same way: compartments for passenger cabin baggage, stationary lighting units and gasper fans (Fig. 2.3.11).

During flight, a chair is very important for a passenger. It is created on the basis of ergonomics and regulatory requirements. Comfort is often the most important factor for a definite type of traveller. Sometimes passengers spend two-thirds of the day travelling, and in this case chair design must emphasize comfort—the alternative is likely to produce the misery of back pain. For passenger chair modeling the Polygon Extrude method is used.

In Fig. 2.3.12 the successive steps of chair modeling are shown. This paper covers several versions of a passenger chair configuration subject to the cabin layout comfort level. Fig. 2.3.12 represents successive steps of a chair modeling for the cabin version with face-to-face seats arrangement. The chair model is further elaborated and improved, and at the final stage it is smoothed with modifier Smooth (Fig. 2.3.13).

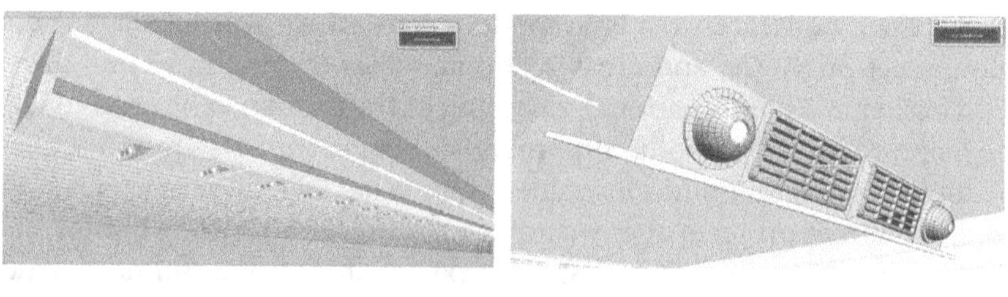

Fig. 2.3.11 Lighting units and gasper fans.

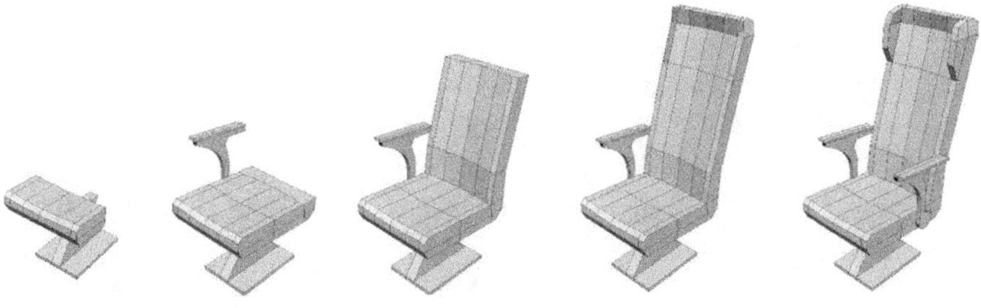

Fig. 2.3.12 Chair modeling successive steps by means of Polygon Extrude method.

2.3.4 Shading of Aircraft Cabin Objects

Before final rendering of the scene it is necessary to prepare objects for shading, create corresponding texture and material. For more detailed shading we divide chair model into component parts by means of command Element (Fig. 2.3.13). Further we create material for chair upholstery, assign it to corresponding elements of the model. As for style the chair upholstery consists of two colors. Bottom, back and head cushion are alike in colour, and the rest parts are of another color (Fig. 2.3.14).

For correct mapping of upholstery texture we apply modifier UVW Map. We apply this modifier in mode Box, that enables to stretch and compress texture in three planes. We carry out similar actions with the second chair and further we need to multiply and arrange them over the cabin.

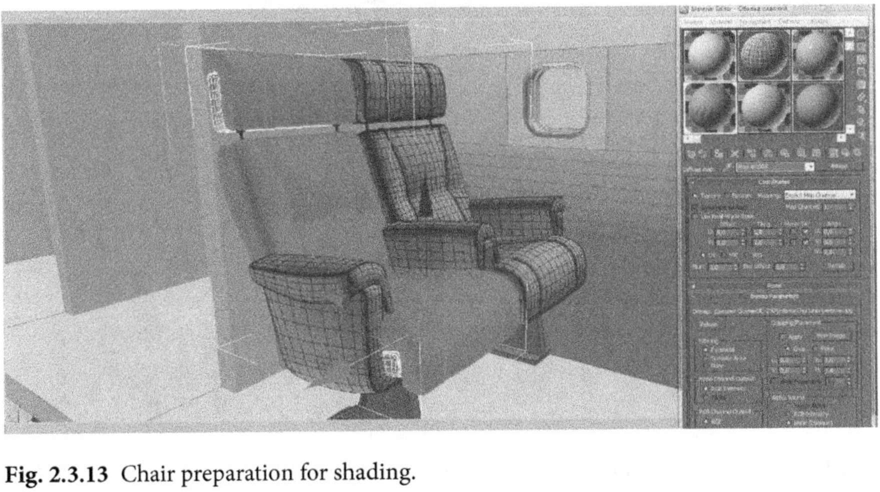

Fig. 2.3.13 Chair preparation for shading.

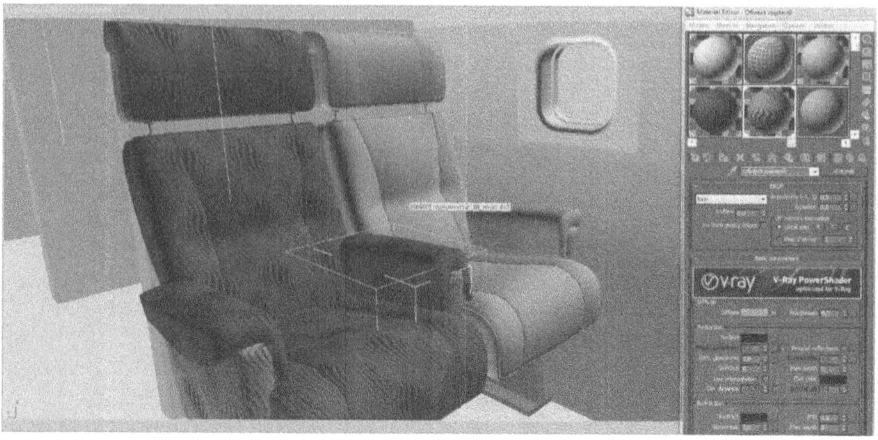

Fig. 2.3.14 Texture assignment to chair upholstery.

Fig. 2.3.15 shows a passenger chair configuration for a comfortable cabin layout. The upholstery of this chair combines two different materials, whose appropriation is made at the elements level.

At the next stage, we distribute passenger chairs all over the cabin along the longitudinal axis in accordance with the layout diagram. For this purpose the graphic system 3ds Max has a convenient command Array. This command enables the assignment of offset value of the next copies relative to each other in all three planes simultaneously. We define optimal distance

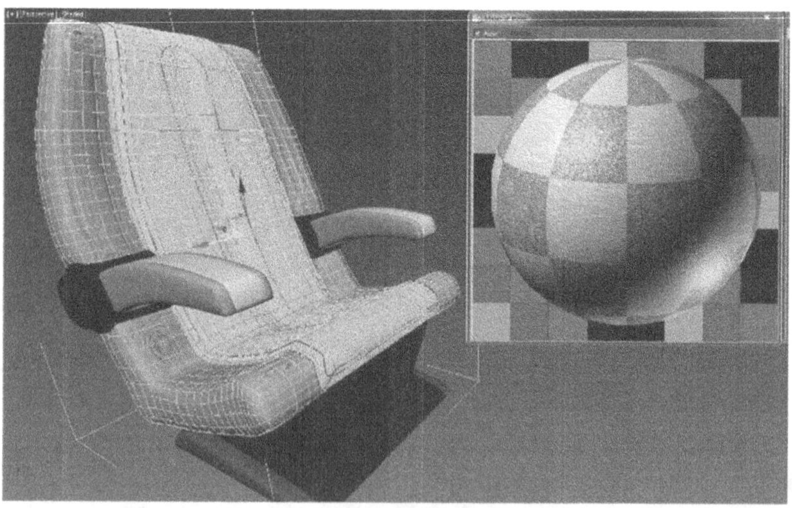

Fig. 2.3.15 Texture assignment to comfortable chair upholstery.

between chairs in accordance with layout drawings selected. Further we assign materials to the rest of the objects of the aircraft cabin interior.

2.3.5 Rendering of Aircraft Cabin

When creating a realistic model, lighting is a rather important detail. For lighting, usually natural or artificial light sources are used. In graphic system 3ds Max there are some rendering methods. However, external rendering module V-Ray creates reasonably realistic renderings. It is based on indirect light method when objects are illuminated diffusely with reflected light. In Fig. 2.3.16, 17 conceptual rendering of amphibian aircraft Be-200 different cabin variants are presented: coupe-type configuration and configuration of enhanced comfort with separate berths.

Fig. 2.3.16 Conceptual rendering of coupe configuration aircraft cabin.

Fig. 2.3.17 Conceptual rendering of amphibian aircraft Be-200 cabin of enhanced comfort with separate berths.

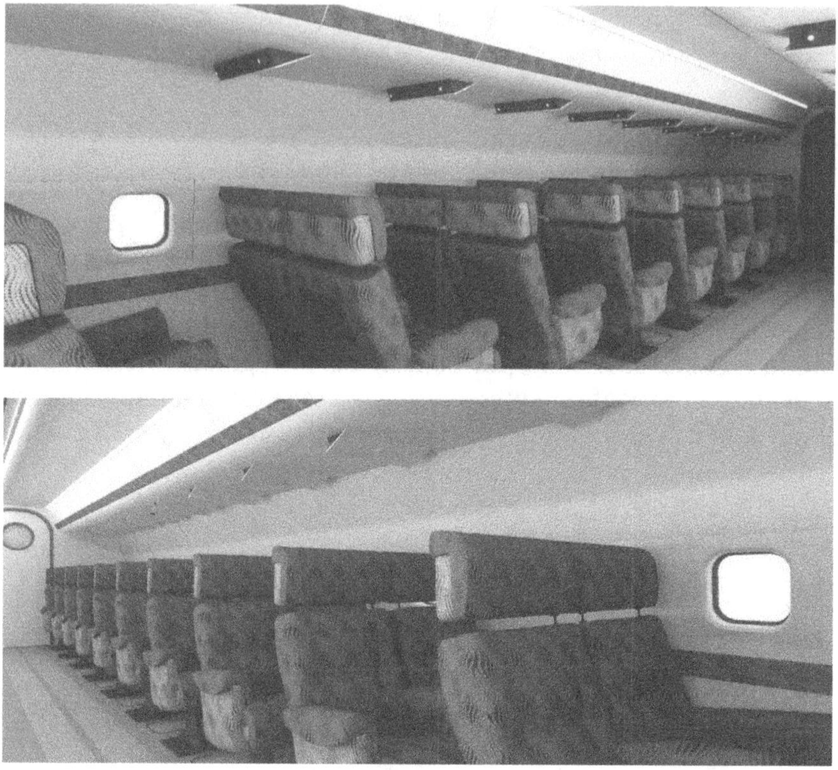

Fig. 2.3.18 Rendering of economic variant of amphibian aircraft Be-200 passenger cabin.

In Fig. 2.3.18 realistic rendering scenes of economic variant of amphibian aircraft Be-200 passenger cabin layout are shown.

Fig. 2.3.19 represents the realistic rendering scenes of Be-200 amphibian passenger cabin comfortable layout (based on the scheme in Fig. 2.3.7).

2.3.6 Conclusion

Due to developments carried out it ought to be noted that for current structure of amphibian aircraft Be-200 different philosophies of passenger cabin layout were developed. Herewith the authors suggest new original concepts of passenger cabin layout intended for various purposes. The suggested concepts have been not only designed, but developed to the 3D-model with realistic rendering. 3D modeling of interior objects was

Fig. 2.3.19 Rendering of comfortable variant of amphibian aircraft Be-200 passenger cabin.

carried out on the basis of well-known methods. The given scenes of passenger cabin realistic rendering give visual presentation about the future of the industrial product. Modern tools of computational modeling make it possible for designers to examine projects of future products in different variants until their real creation.

2.4 Computational Modeling of Amphibious Aircraft Be-103

2.4.1 Introduction

Hydroaviation is effectively used at water areas located all over the world that are hard to reach by other means of transport. An amphibious aircraft may use either ground runways and inland water bodies or offshore zones for takeoff and landing. This feature belongs to the universal qualities of amphibious aircraft. This work is dedicated to aspects of Be-103 amphibious aircraft 3D computational modeling. This aircraft was developed by Beriyev Aircraft Company.

Modeling state-of-the-art technologies are traditionally applied for aircraft designing. The Paper [14] describes the parametric designing of aircraft aerofoil surfaces with the software based on smooth surfaces.

The paper [7] is dedicated to aspects of high-precision software tools to be implemented to optimize the design process when developing an aircraft design. The paper [13] describes particularities of the aircraft design based on the lofting method. Quite a detailed description of modern automated design systems is given in the book [29]. Particularities of preliminary and conceptual aircraft design are given as well.

Designing of Be-103, the light amphibious aircraft, began in the early 1990s at Beriev Aircraft Company [2] (Fig. 2.4.1). The main objective of the project was to create multipurpose amphibious aircraft for local airlines. A low skimming foil was the distinctive feature of this project. Such configuration gave an essential benefit in stability while moving on the water during the takeoff and landing, and increase of the wing lift force due to the screen effect. Through other analogs found worldwide we can mark Airmaster Avalon-680, an amphibious flying boat produced in the

Fig. 2.4.1 Be-103 amphibious aircraft.

USA, and Do-24, an amphibious aircraft produced by German company Dornier Seastar [30].

Be-103, the light, multipurpose amphibious aircraft, may be effectively used in coastal zones, on rivers and lakes. The aircraft is a low-wing monoplane that has vertical and horizontal tail plane and tricycle landing gear with a nose wheel. To ensure its floodability, the boat and the wing of the aircraft are divided into compartments by watertight partitions. The landing gear compartments are also separated from the internal case of the boat by watertight walls.

The power plant consists of two piston engines mounted on the horizontal pylons on both sides of the body. The low-wing scheme creates a considerable screen effect while taking off and landing, and allows the gliding on the step and trailing edges, located in the centre of the wing, at the same time.

Aircraft sea worthiness is sufficient to withstand the roughness up to two points (wave height to 0.5 m). The aircraft can be operated on bodies of water as long as their depth is not less than 1.5 m and the length is not less than 600 m, and on the ground airfields. Maximum takeoff weight of Be-103 amphibious aircraft is 2270 kg, maximum cruise speed is 240 km/h [2]. The aircraft can take 4-5 passengers or cargo of 385 kg. In addition to passenger-and-freight configuration the following options are available: sanitary, environmental monitoring of water areas, fire-prevention monitoring of forest, agricultural.

2.4.2 Modeling Methods and Stages

To develop the 3D-model of amphibious aircraft you can use various software products and modeling methods. In this work we will use capabilities of the 3D graphic modeling system, 3ds Max. The graphic system, 3ds Max, is a type of versatile tool that can interact with many CAD systems [16].

To describe the complete three-dimensional simulation procedure, the flow chart of the main design stages is provided in Fig. 2.4.3:

- preparation of initial projections drawings (Fig. 2.4.4);
- development of three-dimensional projection planes;
- stage-by-stage modeling of the structural parts of the amphibian aircraft;
- assembly of the final three-dimensional model;
- creating and assignment of the materials and textures (shading);
- selection and installation of the light sources;
- rendering of the overall scene.

To create the preliminary design, a set of graphic information should be available in the form of photos (Fig. 2.4.1) and drawings of object modeled (Fig. 2.4.2). Be-103 amphibious aircraft has the following overall dimensions: wingspan = 12.72 m, length = 10.56 m, height of 3.76 m. The processes of 3D-modelling starts from creation of three perpendicular planes, then the images of projection drawings (Fig. 2.4.3) are placed in these planes [18, 31]. This method was tested during the computational modeling of Be-200 amphibious aircraft [21].

To create a model of the body surface, we will use a method of polygonal extrusion. We will also create a half of the model along the longitudinal axis, taking into consideration the object's symmetry; this will simplify the process. In this case we will create the right half of the model; the left one will be made automatically at the final stage.

At the first stage of modeling you should create an initial polygon. The polygon then becomes a starting point for the entire body. Further you

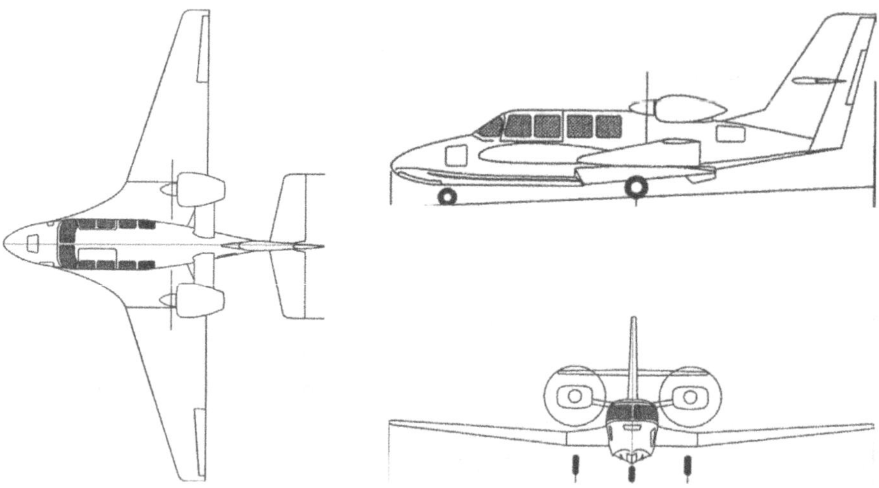

Fig. 2.4.2 Be-103 orthogonal projections.

Fig. 2.4.3 Flow chart of the main design stages.

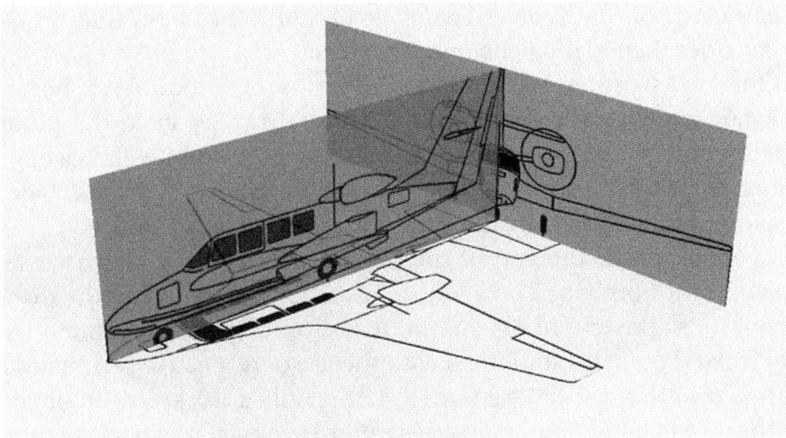

Fig. 2.4.4 Creation of three mutually perpendicular planes.

should consistently replicate a side of the polygon; this stage is carried out by repeated visible body projection. The result of construction is shown in the Fig. 2.4.5. While extruding the sides it is important to keep the constant number of polygons along the body to avoid further problems related with geometry and subsequent finishing of the model. Therefore at the initial stage we must specify the minimum quantity of polygon sides, because

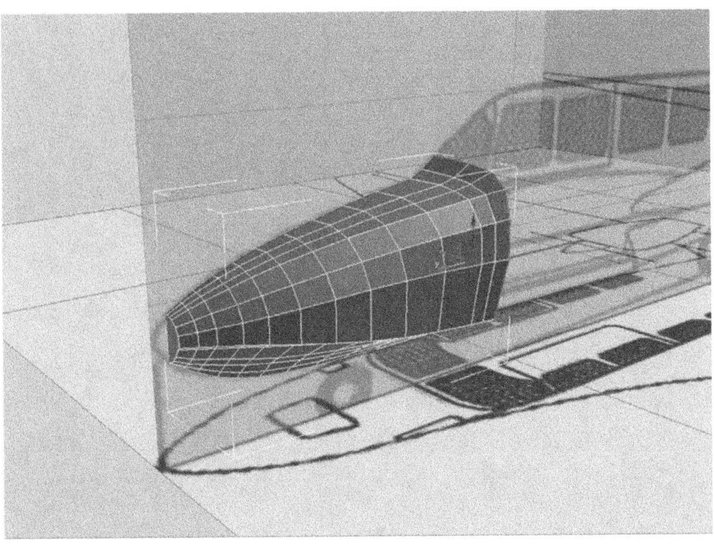

Fig. 2.4.5 Creation of initial polygon in the forebode.

it is easier to add the required number of additional sides than to delete existent sides if an inadequate number is met.

At the next stage we will replicate the contours of the aircraft body (see Fig. 2.4.6) applying the method of sequential extrusion to the group of polygons and subsequent projections alignment. While increasing the number of polygons moving along the body axis, you must collate the extrusion process with aircraft projections in all three planes.

As you approach the end of the body, a grid density increases, keeping a constant number of polygons in a section but reducing the distance between tops. The end of the afterbody is closed by the polygon, then the initial faceted model is smoothed with the use of respective command. The obtained model is not quite a body yet; it is only a workpiece to obtain all the other parts and components thereafter by polygon extruding method (see Fig. 2.4.7).

Before selecting a number of polygons in the tail unit, it is required to review the aircraft projection from above (see Fig. 2.4.8, on the left). We will use these polygons as initial ones to make the vertical tail plane of future aircraft. The vertical plane consists of two main components, the leading and trailing edge and it is a rudder at the same time. The tail unit is simulated with the use of the polygonal extrusion method according to aircraft projections (see Fig. 2.4.8).

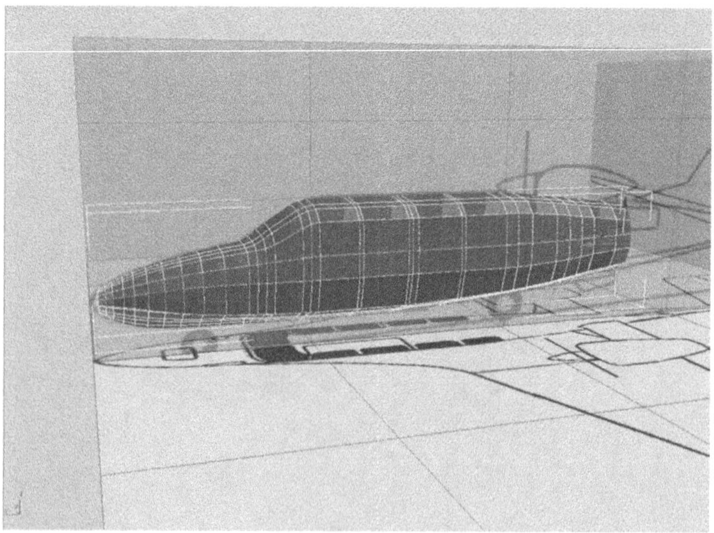

Fig. 2.4.6 Sequential body extrusion.

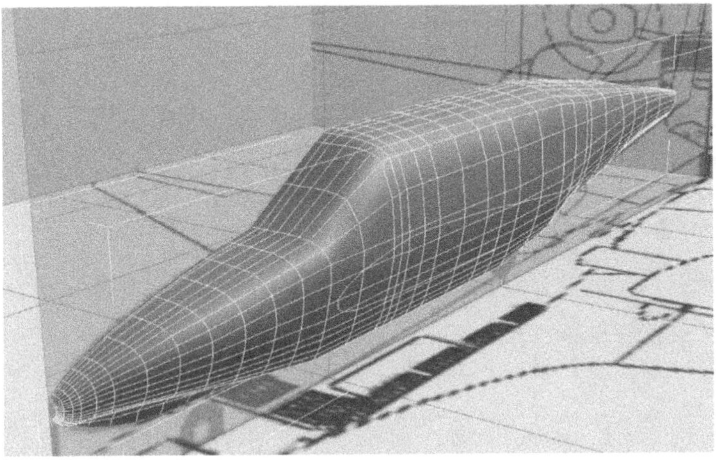

Fig. 2.4.7 Result body extrusion after smoothing.

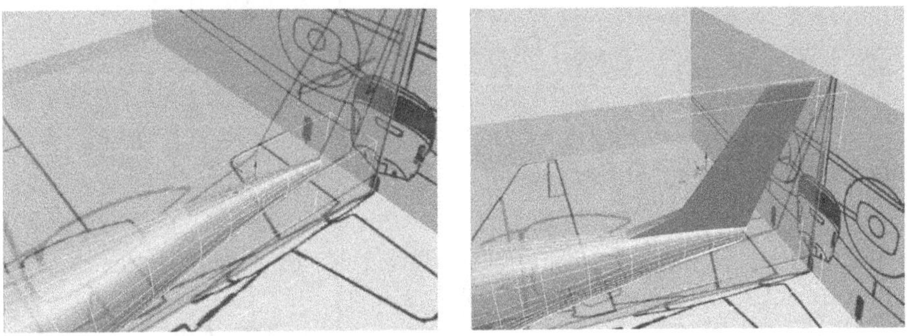

Fig. 2.4.8 Sequential extrusion of tail unit.

Cut the horizontal stabilizer mounting socket contour in the certain point of the tail unit plane with Cut tool (see Fig. 2.4.9). Then you can start sequential polygon extrusion; the result is shown in Fig. 2.4.9.

When the tail unit has been simulated, we begin the simulation of wing, another important part of the future aircraft. The wing of the aircraft has a complex profile as it carries the aircraft while gliding, and acts as the screen, increasing the lift force at the time of taking off from water surface.

The future wing is simulated from the body. The modeling procedure is similar to one used for modeling of the tail unit. Select a number of polygons on the lateral surface to replicate the contours of the wing mounting pad (see Fig. 2.4.10). As a result we will get the polygons used for subsequent

Fig. 2.4.9 Sequential extrusion of horizontal stabilizer.

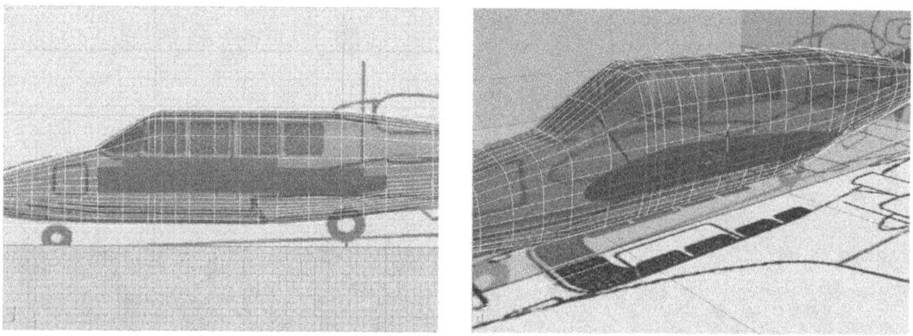

Fig. 2.4.10 Consecutive extrusion of wing.

wing extrusion. Extrude the wing surface the same way, replicating the contour of the wing projection (see Fig. 2.4.11).

All the remained aircraft components, e.g., engine (Fig. 2.4.12), the rotor blades, and landing gear are created the same way.

At the following stage the model geometry is finished and the aircraft is assembled from its components: body with wing and tail unit, engine pylon (strut), engine cover and rotor (Fig. 2.4.13) [19, 20]. Moreover, there are illumination lights on the wing and tail. Detailing of the body implies modeling of a wind screen and side windows. The wing together with rudder and horizontal stabilizer is to be detailed more precisely as well.

In the end of the modeling stage, you must reflect the half of the plane relative to the longitudinal axis (Fig. 2.4.14).

2.4.3 Shading of 3D-Model

To get real photographic images in graphic-oriented system 3ds Max, the correct selection of materials and customization of its main properties are

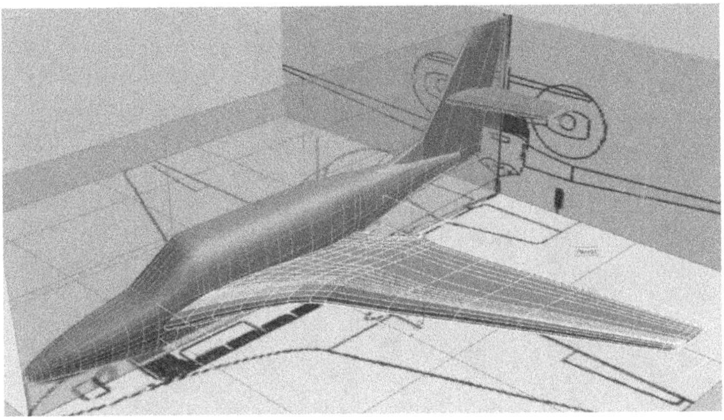

Fig. 2.4.11 Aircraft body with tail unit and wing.

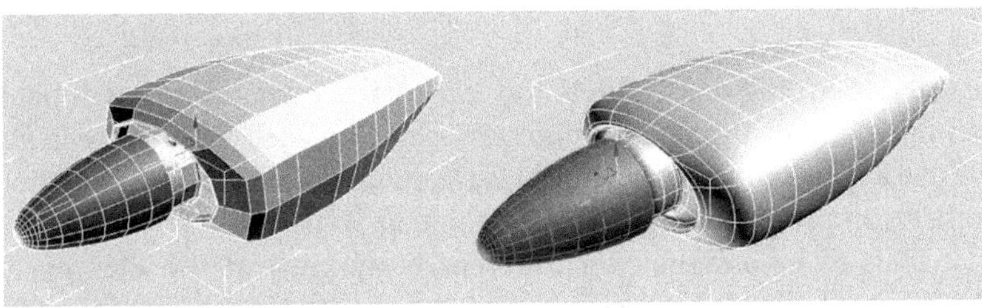

Fig. 2.4.12 Creation of engine cover.

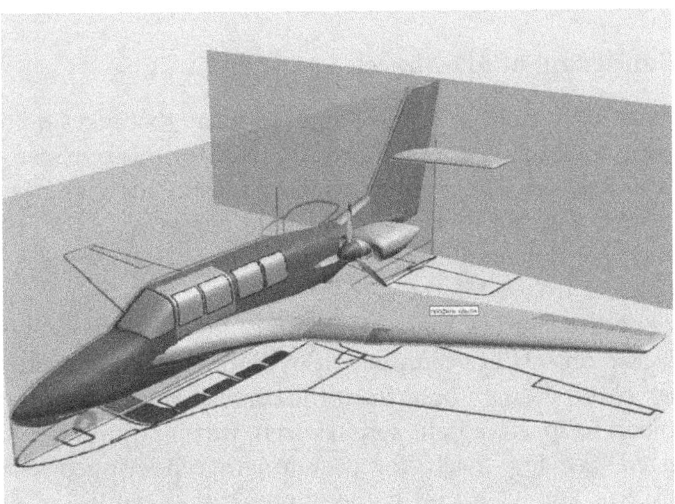

Fig. 2.4.13 Connection of engine cover with the rotor blades.

Fig. 2.4.14 Final assembly of Be-103 model.

to be the depending ones. The V-Ray external rendering unit will be used for rendering, and the use of materials of VRayMtl-type library is stipulated. The key parameter of material is the main color of it (diffusive color); the reflection properties are to be specified as well.

Materials to be assigned for different body parts at the Element sublevel of Editable Poly object (Fig. 2.4.15, on the left). Elements are selected one-by-one and assigned the previously made and adjusted materials with Editor Material tool (Fig. 2.4.15, on the right).

2.4.4 Rendering of 3D-Model

Lighting is an important process of any 3D-scene that takes quite a lot of time. Commonly two types of lighting are used for this purpose – natural and artificial lighting. The graphic-oriented system 3ds Max has standard rendering modules, but their capabilities do not provide the real photographic image. V-Ray module provides the better image. Sources imitating indirect lighting are used in the scene.

In a scene the sources imitating indirect lighting are established. In this mode the objects are illuminated by light that is diffusely reflected from other objects; direct light from the direct light source does not fall on the object. Respective light sources were further applied to the scene and adjusted. See Fig. 2.4.16 for preliminary rendering of the aircraft 3D-model. See Fig. 17 a and b for photographic rendering of Be-103 shaded model.

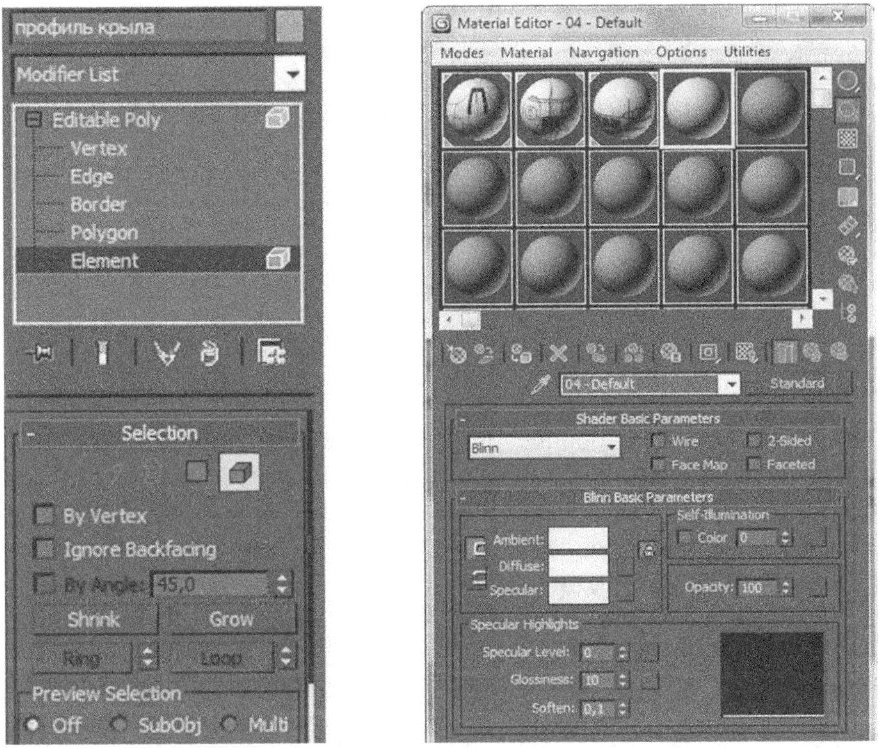

Fig. 2.4.15 Materials assignment on the Element sub-level.

Fig. 2.4.16 Preliminary rendering of Be-103 3D model.

Q6 **Fig. 2.4.17** Rendering of Be-103 shaded model.

2.4.5 Conclusion

Aspects of Be-103 amphibious aircraft 3D computation modeling were reviewed herein. An amphibious aircraft has essential advantages when used in extended coastal water areas, internal reservoirs. No doubt that advanced design and modeling of amphibious aircrafts is an actual target in the development of hydroaviation.

In our case state-of-the-art graphic-oriented systems of 3D computation modeling were used for modeling. Choice of specific modeling method and respective instruments is mainly dependent on the designer's preferences. Resuming the performed developments, we may note that the rendering scenes of Be-103 amphibious aircraft 3D modeling are considerably realistic and they provide visualization of the object designed.

2.5 Conceptual Model of "Lapwing" Amphibious Aircraft

2.5.1 Introduction

Today hydroaviation is actively used in different fields, starting from fire-fighting and effective-rescue operations up to passenger traffic. The issues of applying modern technologies of modeling for aircraft designing are challenging. The most important stage is the development of a preliminary concept of transportation means. Let us review some of the modern literary sources in this field. The article [11] is dedicated to conceptual designing of aircraft, where aerodynamic properties of bird wings are considered. The works [9, 10] study the issues of designing economical passenger aircraft.

The article [32] is dedicated to conceptual designing of passenger aircraft of "flying wing" type. There provided and analyzed are the different variants of aerodynamic configurations. The work [33] contains the peculiarities of conceptual designing of a new generation of supersonic aircraft with original arrangement of landing gear and fuel tank. The article [7] describes the peculiarities of implementing modern program tools for the purposes of designing. There described are the possibilities of a new program for aircraft structure development. The issues of conceptual designing initial stage are described in detail in book [34]. There provided is the methodological basis of idea generation stages and determination of initial requirements for future structure. The book [29] contains the peculiarities of preliminary and conceptual designing of aircraft. Modern systems of automated designing are described in detail.

This work is dedicated to three-dimensional computer-aided modeling of new concepts of amphibious aircraft. It is supposed that the developed model will be in the middle segment of the hydroaviation market. As a result of an amphibious aircraft market review we can remark the following aircraft of low passenger capacity up to 25 persons: Be-103 produced by Beriev Aircraft Company [website of Beriev, 2016], flying amphibious boat Airmaster Avalon-680, produced in the USA, and amphibious aircraft Do-24, produced by German company Dornier Seastar [35]. For the developed model the crew will consist of 2 persons; the passenger compartment can contain up to 24 passengers.

It is necessary to note that the issues of computer-aided modeling of aircraft were studied by the authors in the works [18, 21]. The work [36] provides conceptual visual and graphical solutions of new aircraft based on bionical forms analysis.

2.5.2 Concept Development

In every field of our life, everything in our environment is the product of human thought. The manufacturing of these subjects and objects starts from concept development, the creation of a prototype of the future item [27, 37]. If earlier rather large expenses and materials were required for this purpose, then today in the era of computer-aided technologies this task is simplified; there is no limit for the designers' ideas and imagination.

The process of conceptual development and modeling of transportation means takes several stages. At the first stage the sketch is created, the general view of the future model is drawn: compositional solution; proportion of component parts relative to each other; main style solutions [38, 39].

Based on the analysis of natural shapes rendering, the concept of future prototype is selected.

In the course of concept development there used is the method of designing based on bionical forms. Mammals, fish and birds can provide the designer with interesting visual solutions. At that aircraft fuselage, and mainly the flying boat one, shall meet the requirements of aero- and hydrodynamics at the same time. That is why the designers have the task of searching for a compromise. In the course of a creative search of aircraft outlines some visual and graphical solutions were found, the base of which became natural biological forms living in this environment (Fig. 2.5.1-4).

Based on the analysis of natural forms rendering of off-shore strips the bird lapwing (northern lapwing) was selected. The northern lapwing (*vanellus vanellus* in Latin) is a small bird of the dotterel family; it lives in water ponds and has good flying properties; during mating season the males attract the females by air games (Fig. 2.5.5) [40]. The black-and-white color of its coat will be used for three-dimensional model shading in future. Fig. 2.5.6 provides preliminary design, sketches of the future item forms.

Then the model is drawn in detail with reference to medium dimensions, in which the item is planned for operation, biometric parameters of a man considering the requirements of ergonomics (Fig. 2.5.7, 8) [41, 42].

Fig. 2.5.1 Blue whale and sketch of amphibious aircraft fuselage.

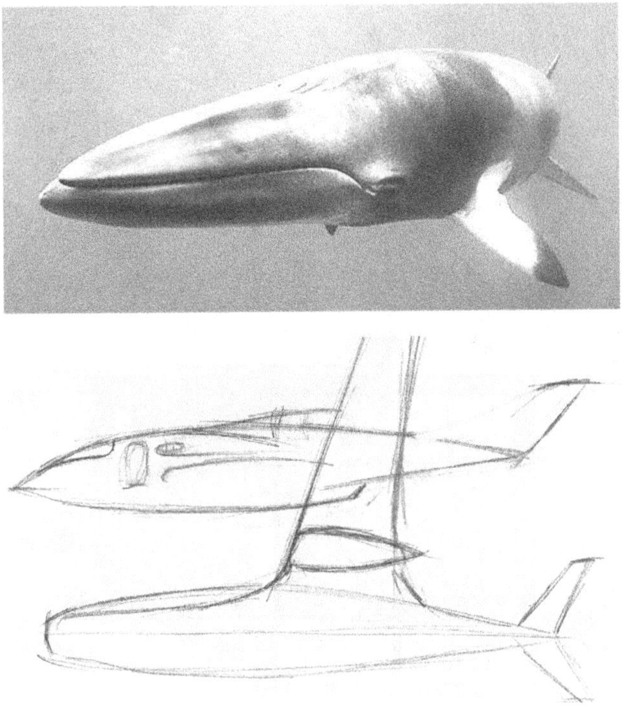

Fig. 2.5.2 Finback whale and sketch of amphibious aircraft fuselage.

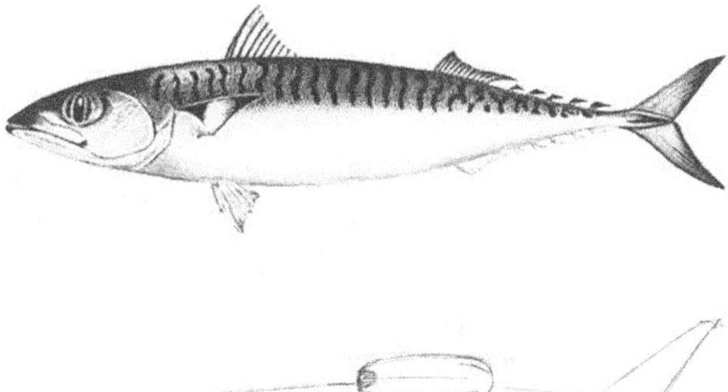

Fig. 2.5.3 Mackerel and sketch of amphibious aircraft fuselage.

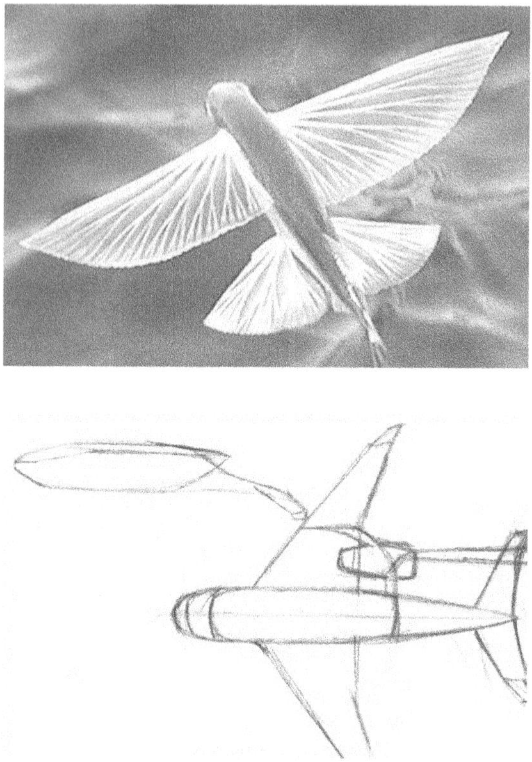

Fig. 2.5.4 "Flying fish" in natural living conditions and sketch of amphibious aircraft fuselage.

Fig. 2.5.5 Lapwing bird.

Fig. 2.5.6 Preliminary sketch.

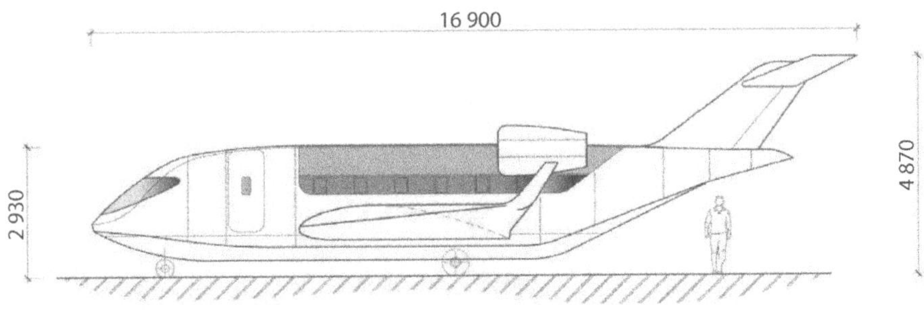

Fig. 2.5.7 Left board view of prototype.

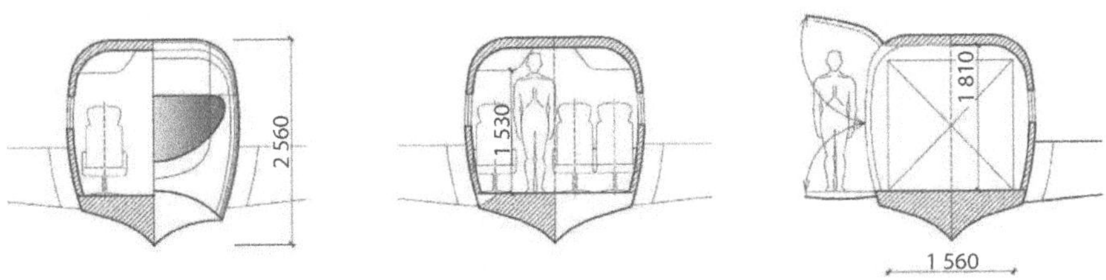

Fig. 2.5.8 Prototype reference to anthropometric and ergonomic requirements.

The base of future hydro-aircraft "Lapwing" concept is a water-borne wing capable of glissading on three points (step, left and right rear edges of center wing). Such a scheme is very advantageous for stable movement on the water at takeoff and landing regimes and increase of seaworthiness. Low location of the wing relative to the boat creates increase of elevating force due to ground effect at takeoff and landing, allows simplifying and lightening the structure of the aircraft (Fig. 2.5.9).

The dimensions of the prototype body shall consider the requirements of future interior and tasks on cargo containers arrangement. Wing span is 18.5 m, aircraft length is 16.9 m, height is 4.87 m.

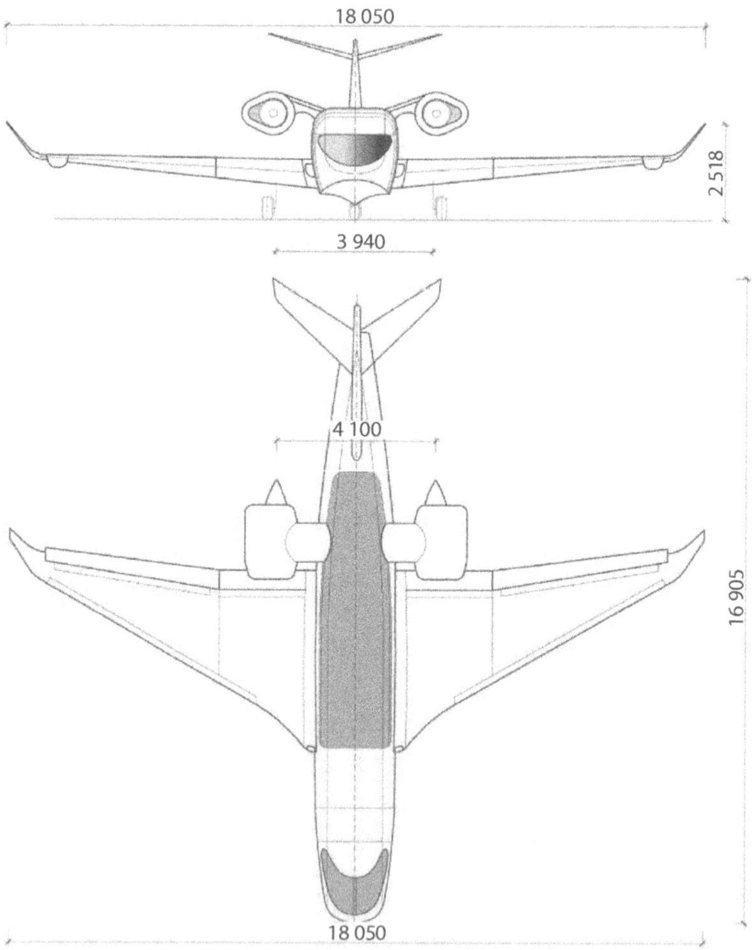

Fig. 2.5.9 Front view and top view of amphibious aircraft prototype.

Fuselage structure can be done from aluminum alloys with the application of composite materials. In the top part of the fuselage there are power elements on the base of solar batteries for partial power supply to the aircraft's onboard network. The aircraft wing has all-metal structure; it has trapezoidal shape with root extensions; it consists of a center wing and two removable panels. On the wing end there are winglets and tips that are designed for increasing effective wing span and lifting force. For the provision of resistance to flooding the wing is separated by waterproof partitions to sections.

Vertical tail fins are single-fin, cantilever. In the top part of the fin there is a controllable stabilizing fin. Landing gear is three-leg type, the diameter of the rear leg tires is larger than the front one. The power unit consists of two turbojet engines located on the pylons close to the fuselage tail part.

For the cargo-carrying variant the increase of fuselage length by 1 m is provided with the help of an insert. It aims to locate the cargo door with dimension 1700 x 1700 mm along the right board. The crew consists of 2 persons (as for business class variant one steward is added). The passenger compartment can contain up to 26 passengers; in the cargo-carrying variant 4 LD2 containers are provided.

Fig. 2.5.10 provides a shaded sketch of a three-dimensional model of amphibious aircraft. The aircraft is designed for use at short-distance lines

Fig. 2.5.10 Shaded sketch of three-dimensional model of amphibious aircraft.

in different regions of the world, in regions with a large number of rivers, lakes, and shallow water ponds that are hard-to-reach for other types of transport. It can be used for transportation of passengers, cargoes, fire-fighting supervision, patrolling, ecological control of water areas, provision of emergency medical care, rendering emergency-rescue works, rest and tourism.

2.5.3 3D Modeling of Amphibious Aircraft "Lapwing"

Modeling of amphibious aircraft structure shall be done with the help of a graphic system of three-dimensional modeling – 3ds Max. The graphic system 3ds Max allows working with drawings made in other graphic packages, thus extending the possibilities of the designer [16]. Three-dimensional models of amphibious aircraft can be created by different methods, one of which is the method of polygonal extrude. For this method the modeling starts from creating three perpendicular planes with aircraft projections located on them.

For fuselage modeling there is the polygon based on Plane primitive element with the number of segments at X and Y axes equal to 1. Later this primitive element shall be transferred into Editable Poly object. According to fuselage projection the object surface is created by sequential duplication of one of the polygon planes (Fig. 2.5.11). At that body half is created for construction convenience with consideration of model longitudinal symmetry.

In the course of planes extruding it is necessary to maintain a constant number of polygons along the whole fuselage in order to prevent problems

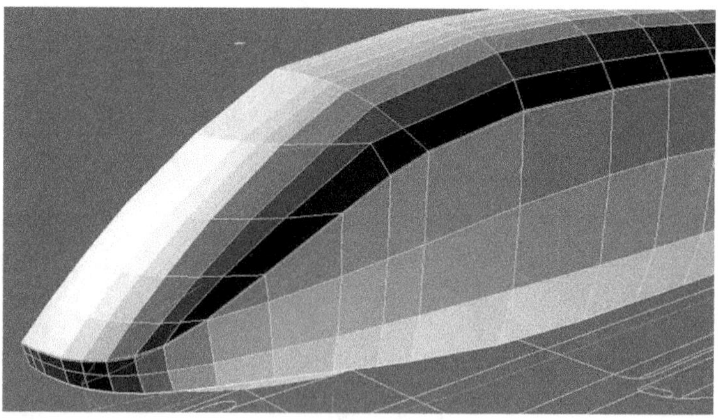

Fig. 2.5.11 Sequential extrude of fuselage polygons.

with geometry and further modification of the model. Then the aircraft body is created by the method of sequential extrusion of a group of polygons followed by projects adjustment (Fig. 2.5.12).

The received result is the base for the fuselage; the other structural parts of the aircraft are extruded by similar method: tail fins, wing, engine pylon, engine body, lifting propeller (Fig. 2.5.13, 14, 15) [19, 20]. The wing has a complicated profile, because it plays the lifting role for the aircraft in glissading mode and works as the screen increasing the lifting force in the moment of takeoff from a water surface.

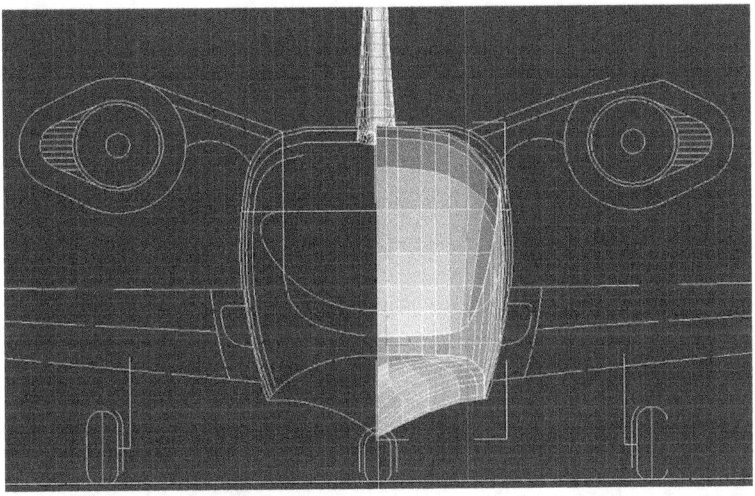

Fig. 2.5.12 Model control in front view.

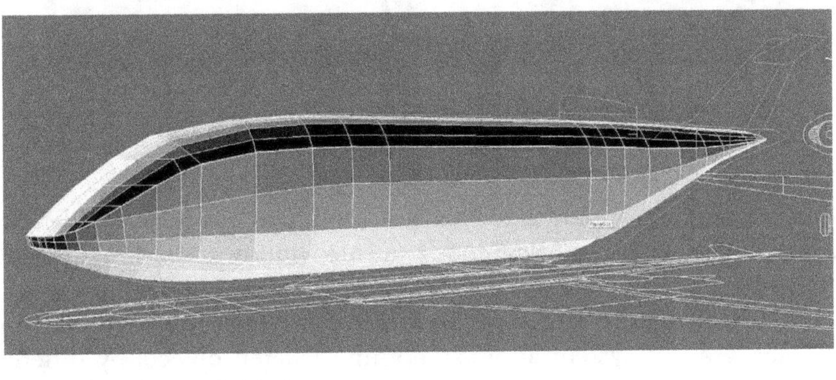

Fig. 2.5.13 Fuselage body blank.

Fig. 2.5.14 Creation of tail fins.

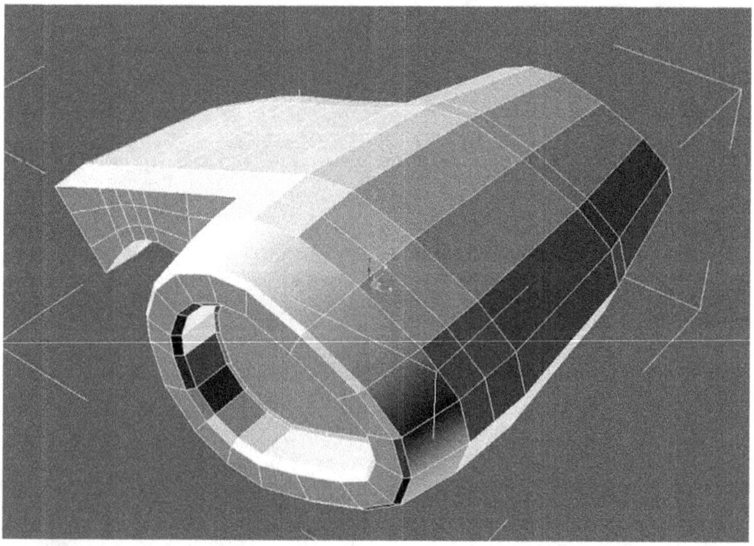

Fig. 2.5.15 Engine body with carrying pylon.

At the next stage, model geometry is modified. Fuselage modification supposes modeling of transparency and side windows. The wing together with steering control and horizontal stabilizer is also designed in detail.

Initially all model component parts are faceted. The capabilities of 3ds Max graphic system allow smoothing faceted objects by different

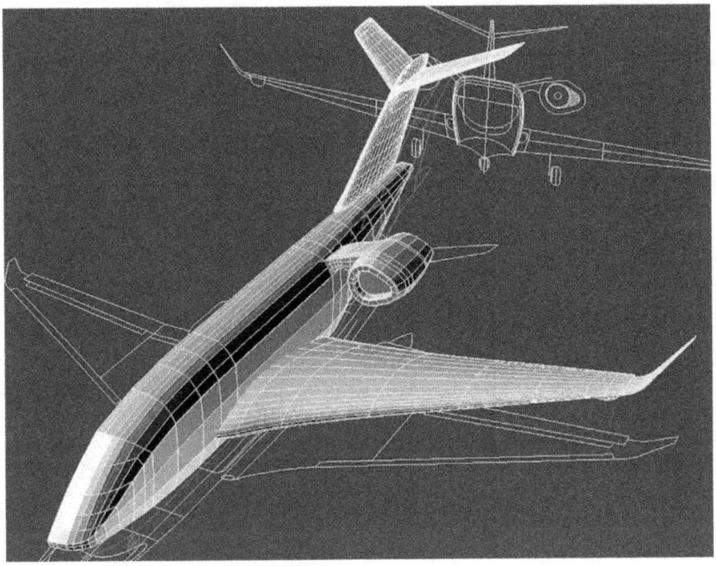

Fig. 2.5.16 Assembly of aircraft body half.

methods. One of the variants is the application of smoothing method NURMS (Non Uniform Rational Mesh Smooth). When surfaces are smoothed the second mirror-like longitudinal half of the aircraft is constructed (Fig. 2.5.16).

2.5.4 Shading and Rendering of 3D Model of "Lapwing" Amphibious Aircraft

The next step of designing is shading and rendering of the constructed model. The process of materials rendering to fuselage separate parts is done at the level of polygons.

After all performed operations we can obtain a finished model for further rendering with the help of realistic models of lighting (Fig. 2.5.17). Integrated V-Ray module is used for scene rendering. Fig. 2.5.18 *a, b, c,* shows final rendering scene of shaded model of "Lapwing" amphibious aircraft.

As a result we can note that the developed three-dimensional conceptual model of amphibious aircraft is performed from creative idea to photorealistic rendering.

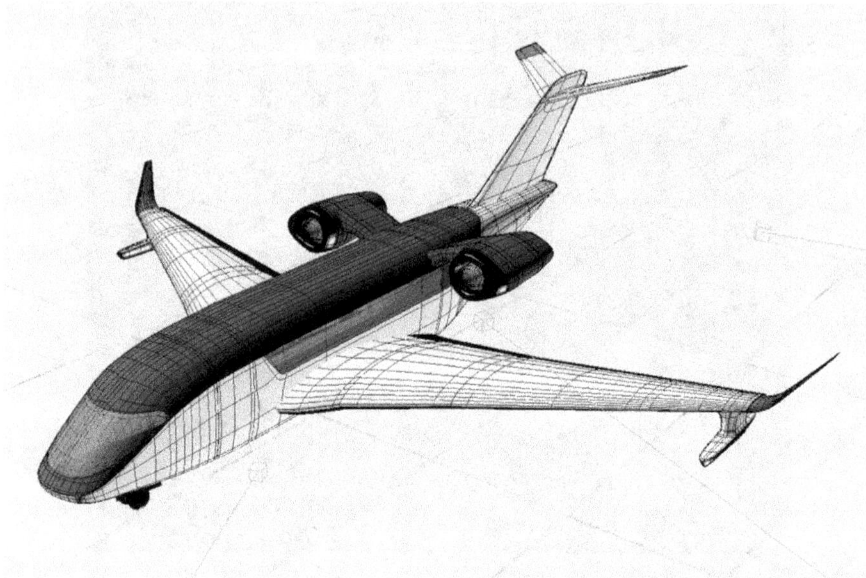

Fig. 2.5.17 Assembled three-dimensional model with rendered materials.

(a)

(b)

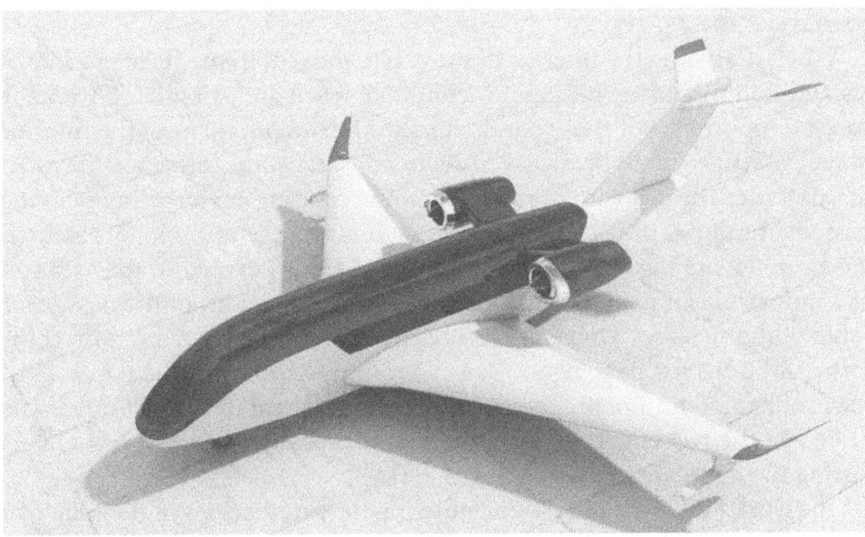

(c)

Fig. 2.5.18 Rendering of "Lapwing" amphibious aircraft conceptual model.

2.6 Computational Modeling of the Cabin Interior of the Conceptual Model of Amphibian Aircraft "Lapwing"

2.6.1 Introduction

At present, the development of tourism in the shelf zones of the World Ocean requires small civil hydroaviation. At the same time there are models of business class aircraft, which were specially designed for private or administrative purposes. The modern aircraft market of this class is represented by a number of airlines that are leaders in their class. Among them, we can mention Dassault Aviation, which has model Dassault Falcon 900 [30]. The Dassault Falcon 900 provides its passengers not only with an enhanced level of comfort, but also with a luxurious interior. Of the Russian aircraft, the Yak-40 model can be noted; along with numerous modifications it has an administrative configuration for the first persons of the state. Small-sized vessels with a capacity of up to 7 people include the Be-103 amphibian aircraft manufactured by the BERIEV Aircraft Company [2]. This light, multipurpose amphibian is designed for use in coastal shelf areas.

When designing the interior of the cabin, the requirements of ergonomics and comfort in conditions of confined space are essential. Given the special importance of these parameters in the development of the interior of the passenger cabin of airplanes, let us consider some references devoted to this topic. The problem of reducing noise, creating a microclimate in the passenger cabin is the problem of current importance, and a lot of research has been devoted to this subject. The work [43] is devoted to the creation of comfortable climatic conditions for the interior design of the passenger cabin of the airplane. The temperature effect and ventilation effect on comfort conditions for passengers was taken into account. In the paper [44], the problem of minimizing noise in the passenger cabin is considered. The design of the passenger compartment and the convenience of passengers during the flight are discussed in [23, 24].

The work [45] is devoted to formulating the peculiarities of the practical development of the cabin interior design, choosing methods, and creating recommendations for the implementation of project design concepts based on hypotheses. The article [25] considers the concepts of the design of the cabins of future passenger aircraft. Various aspects of the optimum configuration of the cabin of the aircraft based on the existing technical design requirements are described. The chapter of the book [26] discusses the development of various configurations of the fuselage pointing out their advantages and disadvantages. Practical recommendations on cabin and cockpit cabin design taking into account the design of the fuselage and ergonomic standards are made. Modern trends in the design of cabins and interiors are presented in detail on the site [46]. The requirements of the design of the layout of the cabin taking into account the comfort and safety of the passengers are presented in work [47].

When designing the interior, the comfort of passengers is often linked to the design of the seat. The article [48] deals with the development of passenger seat construction in accordance with safety and durability standards. The results of the strength testing of a passenger seat under dynamic loads in the airplane cabin are discussed. According to the findings presented in paper [49], the sensation of comfort is mainly associated with passenger experience in the passenger seat. The article [50] describes the features of the design of the passenger seat in the interior design of the aircraft cabin.

The review [51] is devoted to revealing the factors influencing the sensation of comfort and discomfort of passenger seats. According to the conclusions of this review, the relationships between the anthropometric parameters of a human body and positions occupied in the passenger seat are the determining factors. In a recent paper [113], the issues of taking comfort into account on the basis of a passenger survey for the design of a comfortable airplane interior have been considered. The parameters of the passenger's personal space when sitting in the passenger seat have been taken into account as the dominant factors. According to the results of the review of the literature, it can be noted that when designing the interior of a passenger cabin, not only aesthetic qualities are needed, but special attention should be paid to the requirements of ergonomics and comfort.

2.6.2 The Concept of the Amphibian Aircraft "Lapwing"

The conceptual model of the new amphibian aircraft "Lapwing" was presented by the authors in the paper [52]. Based on the bionic forms, visually graphic solutions for the layout of the amphibian aircraft were created. The concept of the amphibian aircraft was named after the lapwing bird, which lives in ponds and makes a virtuosic air game during the mating season. Its black-and-white color of feathering was also used in the shading of the three-dimensional model of the amphibian aircraft.

This paper discusses the conceptual solutions for the layout of the interior of the amphibian aircraft "Lapwing", the step-by-step creation of the passenger compartment from the drawing to the three-dimensional model. The novelty of the article lies in the development of original layouts, taking into account the operational purpose, the creation of an ergonomic model of a new comfortable passenger seat and a cozy interior of the amphibian aircraft cabin. For the passenger modification of the amphibian aircraft Be-200 the interior design issues were considered by the authors in [21].

The intended purpose of the amphibian aircraft "Lapwing" is both administrative exploitation and the transportation of officials of state institutions and commercial organizations. Aircraft of this class have a passenger cabin, which can accommodate from 6 to 10 people. The work of administrative aircraft has a constant character, which helps them to perform operatively. Such aircraft are used by private persons for personal flights, transportation of employees and partners. Today, the term "business jet" is used to refer to aircraft of this class; the word "jet" means only jet engines, but machines with turboprop or piston engines can be used as an administrative aircraft too.

The concept of the layout of the amphibian aircraft "Lapwing" is based on the water-swinging wing with the possibility of gliding at three points (redan, right and left rear edges of the center section). This arrangement has the advantage of the stability of motion on the water at takeoff and landing modes and in the increase of seaworthiness. The low wing position increases the lifting force due to the screen effect on takeoff and landing, which makes it possible to simplify and facilitate the design of the aircraft.

Fig. 2.6.1 shows a sketch of the model of the amphibian aircraft with reference to the operating environment, the wingspan is 18.5 m, the length of the aircraft is 16.9 m, and the height is 4.87 m. The dimensions of the prototype fuselage are designed taking into account the requirements of the future interior and with space for the placement of freight containers. Fig. 2.6.2 shows the prototype drawings of the amphibian aircraft tied to

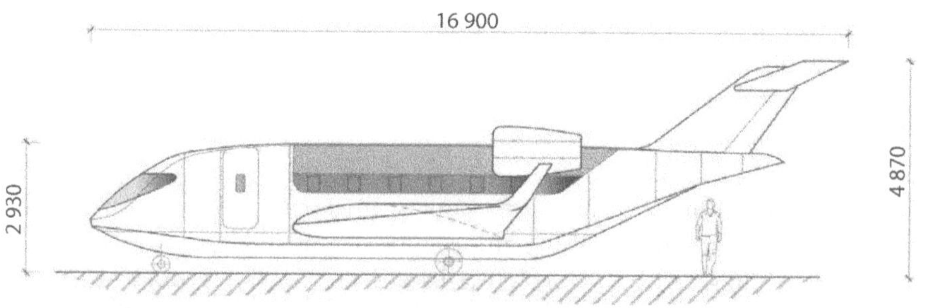

Fig. 2.6.1 View of the prototype from the left side (Fig. 2.5.7.).

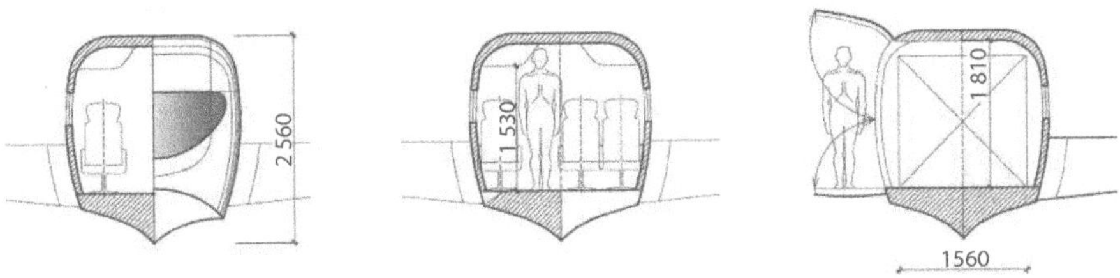

Fig. 2.6.2 Linking the fuselage to anthropometric and ergonomic requirements.

anthropometric parameters, taking into account the requirements of ergonomics [41, 42]. It is proposed that the developing model would occupy the middle segment in the hydroaviation market; the crew will consist of 2 people, and up to 9 passengers can be accommodated in the passenger cabin.

2.6.3 Layout Concepts

The development of any internal space of the aircraft must begin with a new scheme. The design of the cabin of the aircraft, unlike the interior of residential or public premises, is limited primarily by the dimensions of the fuselage and the overall layout of the aircraft. In this case, the designer needs to organize a space that corresponds to all the necessary ergonomic and anthropometric requirements [19, 38, 53]. Along with aesthetics and utility, one should also take into account the hygienic properties of materials for interior design. Trauma safety in flight also depends on the competent ergonomics of the interior of the cabin.

At the stage of searching for the cabin interior concept, you should analyze the layout and arrangement of the equipment, work out several options for layout, depending on the purpose of the aircraft, taking into account ergonomics. Let's consider a number of layout solutions for the interior space of the aircraft, designed to carry 8-10 passengers. The layout of the passenger compartment with passenger seats located in the direction of movement is shown in Fig. 2.6.3; there is a mini-bar with soft drinks in cabin, and a restroom is located in the bow section. Such arrangement of the cabin allows placing comfortable seats with an individual media unit, if necessary; the seats take an ergonomic position for comfortable sleep during the flight.

Lack of time is forcing businessmen to conduct business meetings or briefings, both during parking and in flight. For this purpose, the seats might be placed in face-to-face position with a table between them, for convenience (Fig. 2.6.4). In addition, we can suggest a layout that will divide the cabin of the aircraft into two compartments for individual meetings

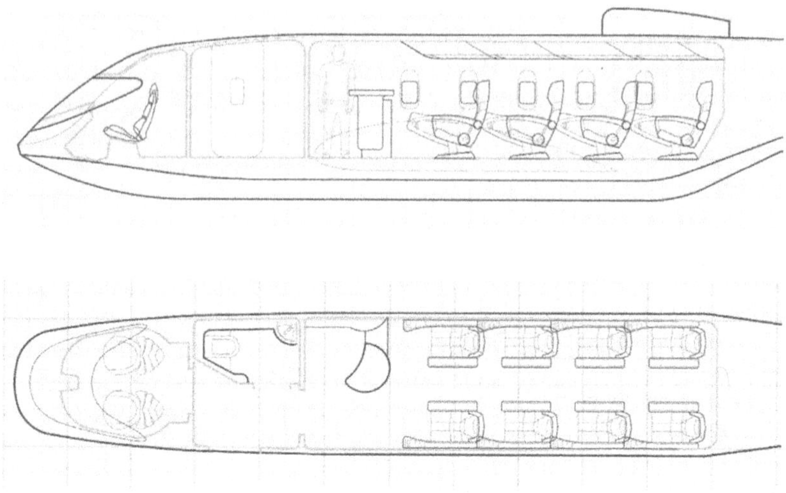

Fig. 2.6.3 Arrangement of the cabin in two types with the placement of seats in the direction of movement.

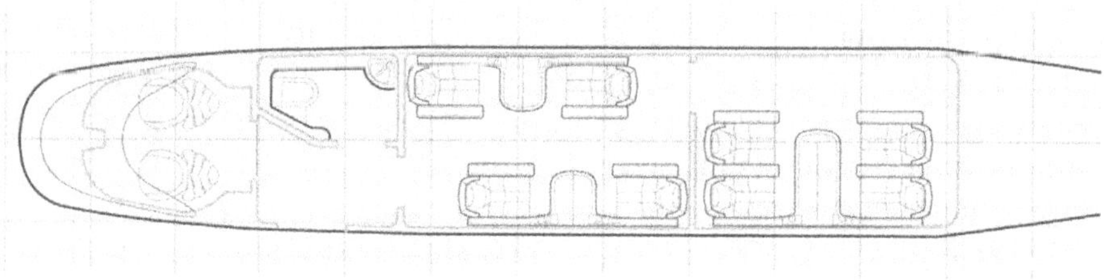

Fig. 2.6.4 The layout of the cabin divided into two compartments for work.

and negotiations. This arrangement will allow, if necessary, the seats to be transformed into sleeping places (Fig. 2.6.5).

The layout solution affects not only placement in horizontal directions, but also vertically. Fuselage contours provide for maximum use of internal space with external compactness [47] (Fig. 2.6.6). In the design process, it is necessary to work out the dimensions of the elements of the passenger compartment and passenger seat, taking into account the anthropometric parameters [50].

2.6.4 Development of a Passenger Seat

Development of the seat begins with a detailed drawing that takes into account ergonomic requirements and human parameters (Fig. 2.6.7), the

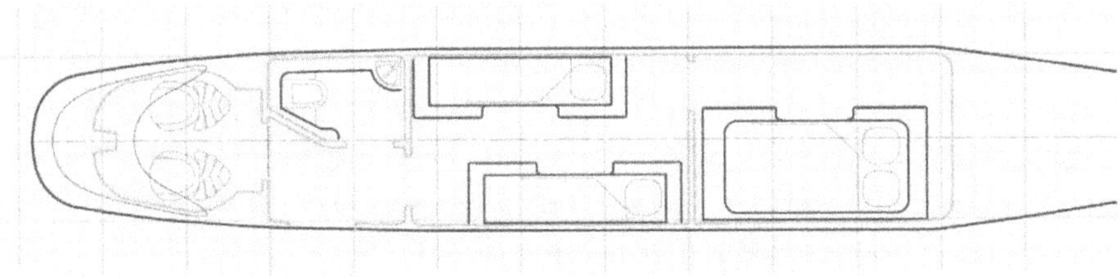

Fig. 2.6.5 Transformation of passenger seats into sleeping places for a cabin with two compartments.

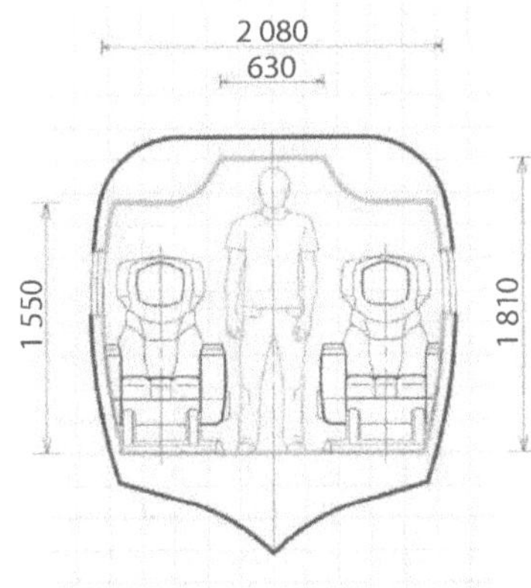

Fig. 2.6.6 A cross-section of the cabin with the main dimensions.

length of the seat is assumed to be 1.258 m, height – 1.211 m, width – 0.732 m. The body of the seat is made in the form of a bucket; the angle of the seat backrest makes it convenient to spend many hours in flight. This drawing will be the basis for three-dimensional modeling of the passenger seat in the future.

For computer modeling of the interior of the amphibian aircraft cabin, the graphic system 3ds Max is used [16]. Modeling of interior objects can be done in several ways, one of which is the method of sequential extrusion of polygons (Polygon Extrude). This method allows you to more accurately track the coordinates of the polygon vertices in the modeling process based on a two-dimensional drawing. As a result of multiple repetition of the Polygon Extrude operation, half of the seat model is created for further processing. The seat has a vertical axis, the axis of symmetry, which allows making a mirror half.

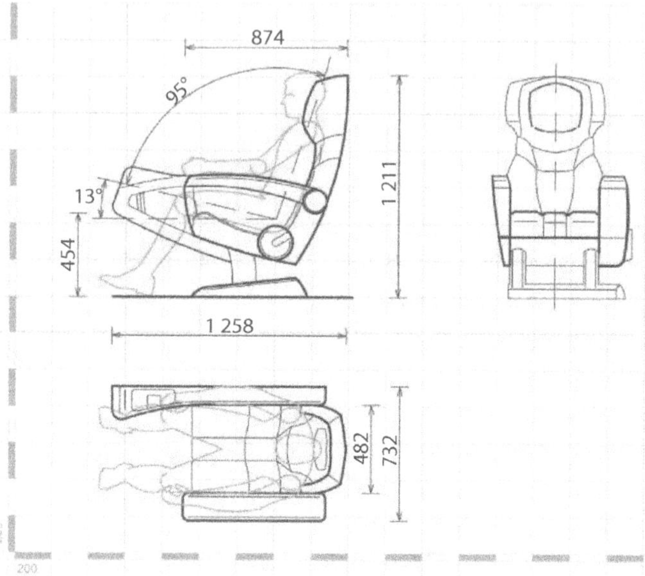

Fig. 2.6.7 Drawing a conceptual model of a passenger seat.

At the next stage, the geometry of the model is being refined; initially it has a faceted surface (Fig. 2.6.8, left). The capabilities of the 3ds Max graphics system make it possible to smooth the model in various ways. One of these methods is to convert the Editable Poly model to objects of the NURMS subdivision type [54]. The result of applying this command

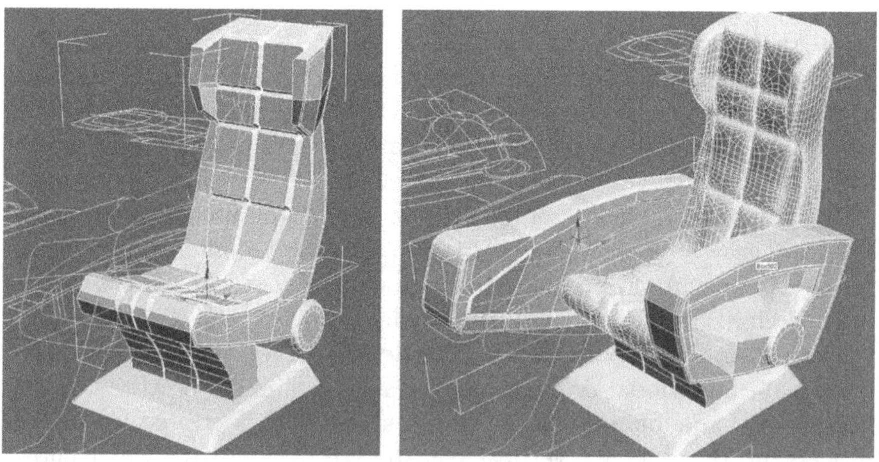

Fig. 2.6.8 Primary faceted model and final assembly of the passenger seat.

significantly improves the appearance of the model (Fig. 2.6.8, right). Before proceeding to the assembly stage, the armrests of the passenger-fat seat are being modified.

2.6.5 Modeling of the Cabin Interior

Next, all other interior objects of the aircraft cabin are created. When making a realistic model of the fuselage, the developed plans, cuts and sections of the future fuselage are used. The main fuselage contours were performed on the basis of the Spline Extrude method.

In the next stage, the seats are distributed along the longitudinal axis in the cabin. For this, the 3ds Max graphics system has a convenient tool called Array. The tool window allows you to set the offset value of subsequent copies relative to each other, and in all three planes simultaneously. At the final stage, all the interior objects of the future model are assembled (Fig. 2.6.9). To maintain the accuracy of the alignment, you should use object snapping, and then you need to go to the materials setup and to the scene lighting setup.

2.6.6 Assignment of Materials and Rendering of the Scene

The objects around us have important properties that determine their appearance. To make photorealistic images, it is essential to make a right choice of material and setting of its basic properties, including lighting.

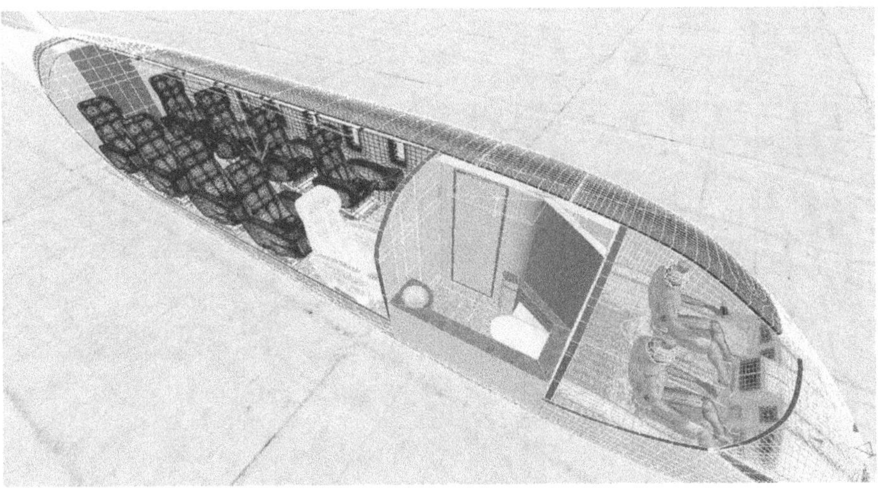

Fig. 2.6.9 Final interior assembly.

These realistic parameters can be provided by the external rendering module V-Ray, so the material library VRayMtl is used to render objects. One of the important parameters of the material is its basic color (diffuse color). To make a realistic simulation of the seat, you need to apply materials to it at the sub-object level. The process of materials application begins with the partitioning of the model at the Element level of the Editable Poly object.

Fig. 2.6.10 shows the process of creating a seat arm material using the material editor. By displaying a sample of material (Fig. 2.6.10, left), it can be determined that the armrests will simulate the surface of the lacquered redwood.

Fig. 2.6.11 shows the scene of rendering of the passenger seat with the assigned materials. Objects of the scene can be visualized with varying degrees of accuracy. The graphics system 3ds Max uses several rendering mechanisms: to examine objects in the projection window, to view the sketches of the materials and to obtain the final image. These mechanisms allow us to find a compromise between the speed of rendering and the quality of the image. Fig. 2.6.12 shows the scene of rendering of the interior of the cabin of amphibian aircraft after the arrangement of seats, and the restroom is shown in the bow part.

The external rendering module V-Ray is based on the creation of indirect lighting; this method is used by many of the modern rendering modules. Indirect illumination implies the illumination of a scene only by diffuse reflected light from other objects, without the participation of a direct ray of light from an immediate source of light. In Fig. 2.6.13 there is the final stage of rendering of the toned interior model of cabin of the amphibian aircraft “Lapwing”.

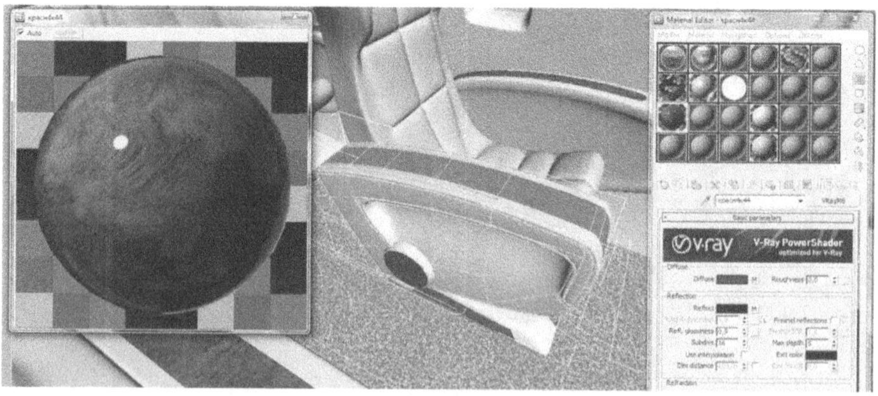

Fig. 2.6.10 The process of assigning material to individual seat model polygons.

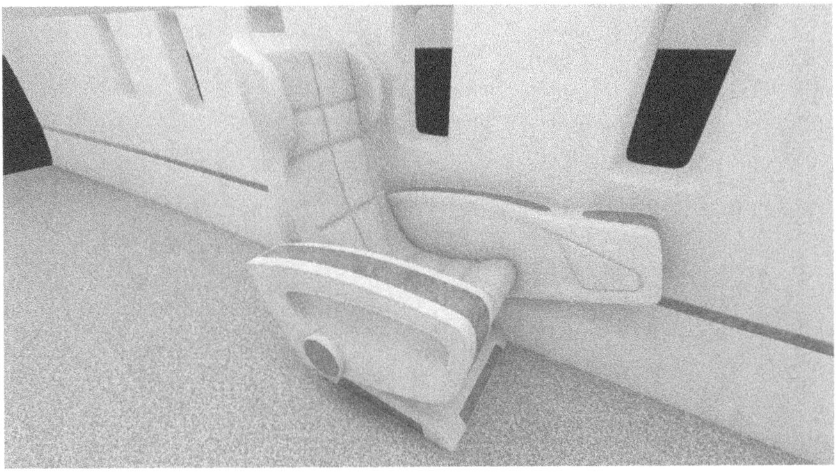

Fig. 2.6.11 Rendering of the conceptual model of the passenger seat with the assigned materials.

Fig. 2.6.12 Final rendering of the cabin on the top view.

2.6.7 Usability and Comfort Cabin Interior

In connection with the special importance of the requirements of usability and comfort, we will describe the proposed convenience criteria for the interior design of the airplane cabin. Each detail of interior finish is aimed at creating maximum comfort for passengers. Thus, the interior surface of the cabin is lined with synthetic incombustible materials of light, soft

Fig. 2.6.13 The final rendering of the passenger compartment of the amphibian aircraft "Lapwing".

tones that do not cause passenger fatigue during the flight and create the impression of spaciousness [24]. Armrests are made with wooden inserts, faced with decorative fiberglass. The floor is covered with a synthetic carpet; the windows are equipped with curtains-light filters. Materials used in the finish meet the requirements of a fire hazard for the interior design of the airplane cabin.

Passenger seats are not only aesthetic and beautiful but also comfortable. The dimensions of the seats, their proportions, as well as the properly selected profile of the seat and the backrest, reduce passenger fatigue during a long flight [49]. Chair backs, as usual, can be deflected back, so a passenger can take a comfortable pose, and in the case of a rough landing, the chairs can also lean forward. This feature of the seat design protects passengers seated in the back from bruises.

The chairs are made in the form of a single-seat block. The blocks of the left and right sides of the cabin are made in a reflected form and have a similar attachment to the floor rails; there are pockets for storing individual lifejackets when flying over water areas and safety-belts. The minimal step of installation of armchairs is 1m. The backrest of the chair in its original position is inclined to back from the vertical by an angle of 15°, and by means of the adjustment mechanism, it can be deflected back from the initial position by another 25° with fixation of the intermediate positions.

Mechanisms for deflecting and fixing backs of the seats are mounted on the extreme armrests. When you press the button, you can deflect the seat back into the desired position; when the button is released, the backrest is fixed in the deflected position. When the button is pressed again, the spring acting on the lever returns the backrest to its original position. The armrests of the chair do not recline; for the convenience of service of the interior, backrests recline on seats, and seat pads rise up to a vertical position. The seat framework, which is the main element of the chair, consists of two pipes, on which legs of the chair and supports for installation of armrests, backrests, seats, and levers of folding tables are mounted.

There are solid shelves for small items along the two sides; on the lower surface of the shelves, there are individual passenger service units, on which adjustable individual ventilation nozzles, adjustable reading lamps, and flight attendant call buttons are mounted. The cabins are well lit by the ceiling lights. At night flights, when the general lighting is off and the telltale light—a light bias behind the top panel—remains, a passenger can switch on one of the three adjustable lamps located above his or her head on the individual passenger service unit.

2.6.8 Conclusion

In conclusion, it can be noted that passenger comfort criteria remain one of the most important parameters when designing an interior of an airplane cabin. While this condition must be met, we should not forget about the space limitations. In this paper, we made an attempt to create a concept for the ergonomic interior of a light aircraft of the business jet class. In this article we proposed original layout schemes for a salon of different operational purposes, an ergonomic model of a comfortable passenger seat, and a cozy interior of the amphibian aircraft cabin was designed. As a result of developments, it can be noted that the resulting photorealistic rendering of the interior of the passenger compartment can visually show the interior design features of the future aircraft.

2.7 Conceptual Model and Interior Design "Water Strider" Ekranoplan

2.7.1 Introduction

Now there is the fissile operation of the continental shelf, areas with stretched water areas, also polar regions with special climatic conditions. For these purposes mobile, efficient vehicles for various purposes, both on water and on land are required. These tasks can be quite solvable by means of vehicles on the basis of wing-in-ground effect. Such ships are called ekranoplan or Wing-in-Ground craft. They fill a technological gap between planes and surface ships. Historically the most developed technology of practical development of these vessels was created in Russia, which explains the use of the name (ekranoplan) [1, 55].

It is possible to point to the advantages of the ekranoplan: high profitability and load-carrying capacity at the expense of a padding body force of wing-in-ground effect; low-visibility because of low flight altitude; use for snow-covered Arctic plains without the need for basing infrastructure, is enough water area or the flat land area. But they have also shortcomings: the control of the ekranoplan demands specific skills from the crew; they have a low maneuverability; it can meet obstacles (birds or shallow vessels) on airway that complicate navigation.

Ekranoplan-amphibians have larger prospects in the field of rescue of people who were injured by disaster at sea. Rescue ekranoplan can splash down, and the whole medical center for ensuring help for the wounded can be placed onboard. Also they can be applied to transportation of passengers, freights, fire-prevention supervision, patrol, environmental control of water areas, rendering of urgent medical care, providing a wrecking, rest and tourism. It can be noted that the leading sea powers planned production of special large ekranoplans for transportation of goods for a long distance on transoceanic open spaces.

In this work the concept of a small-size ekranoplan-amphibian (aerodynamic ground-effect craft) is considered for use on lines of small extent in various regions of the world, in regions with a large number of rivers, lakes, and shallow waters which are remote for other means of transport. The concept includes development, both bionic configuration of contours of the vessel, and an internal interior of passenger salon and basing infrastructure options, beginning from sketches to the finished three-dimensional model. This concept is original, is based on historical experience of creation of ekranoplan with use of the modern technologies of projection and use of perspective materials.

The existing modern small-size ekranoplan use often not only the screen mode; they can also fly at larger heights in the form of the aerodynamic ground-effect craft with properties of amphibians, seaplanes and high-speed boats. Let's consider the review of analogs and the current state of the world market of small-size ekranoplans.

2.7.2 Review of Ekranoplans

The ESKA-1 aerodynamic ground-effect craft (the rescue amphibian boat) (Russia) was developed by the engineer Grunin E.P. in 1973 for rescue operations on water (Fig. 2.7.1) [114]. Aerodynamic configuration uses wing-in-ground effect and is based on A. Lippish's projects. For transversal stability, wings on the ends have the aerodynamic ledges located at an angle and equipped with removable ailerons. Horizontal tail plumage is established behind the wing at the most remote distance. The crew consists of two people, the scope of a wing is 6.90 m, length of 7.80 m, the cruiser speed of 100 km/h, flight altitude in the mode of the screen from 0.3 to 1.5 m. Now the given configuration is often used by modern producers in the different countries [55].

Volga-2 is the first ekranoplan in Russia of civil appointment, developed by designer V. Dementiev (from Alekseev's design bureau); the first flight took place in 1984 (Fig. 2.7.2) [56]. It falls into a class of river crafts on the dynamic airbag. Volga-2 has the streamline housing equipped with an air wing, vertical and horizontal stabilizers with wheels of the direction and pitch, inflatable pneumocylinders – the floats installed from below on a housing and side skegs. It is intended for eight passengers and one crew member, a range of a wing of 8 m, length of 11 m, flight altitude in the mode of the screen of 0.5 - 0.8 m, the cruiser speed of 120 km/h, range up to 500 km. The ekranoplan is designed for high-speed passenger traffic, the fissile tourism, business trips, and rescue operations. Landing of passengers can be made from not equipped coast or the mooring.

Passenger ekranoplan "Aquaglide-5" developed by ATTK-INVEST (Russia) represents a small passenger vehicle. It is intended for transportation of five passengers with a speed up to 170 km/h apart up to 450 km

Fig. 2.7.1 ESKA-1 aerodynamic ground-effect craft.

Fig. 2.7.2 Ekranoplan Volga-2.

(Fig. 2.7.3, at the left) [57]. Length of an ekranoplan is 10.66 m, width of 5.9 m, height of 3.35 m and a housing, consists of a wing with flaps, terminating parts of wings (skegs), the fin with rudders, the stabilizer. As an engine installation, the eight-cylinder petrol engine of Mercedes-Benz is used, the four-blade air screws located on each side can turn concerning the housing on a corner up to 60 degrees. Ranges of application of the ekranoplan: tourism, transportation of passengers, coast guard, rescue operations, environmental control. In Fig. 2.7.3, the interior of passenger salon is also presented on the right.

In 2004 design bureau "Nebo plus more" (Russia) developed the 24-seater "Burevestnik-24" (Petrel-24), aerodynamic ground-effect craft with a useful load of 3.5 t and flying range up to 2000 km (Fig. 2.7.4) [115]. These vessels have three modes of movement; at a speed of 30 km/h the

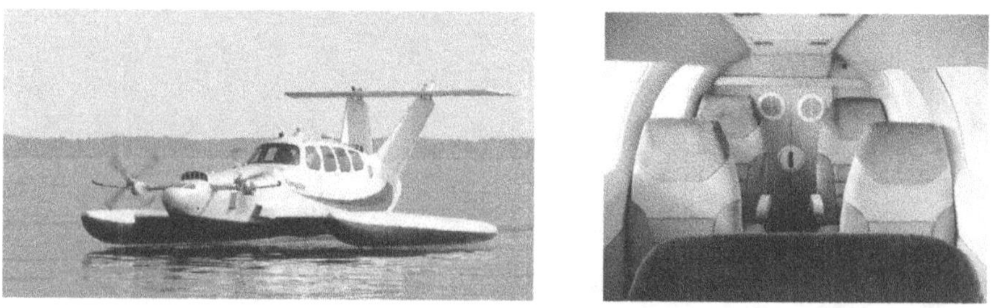

Fig. 2.7.3 Passenger ekranoplan "Aquaglide-5" and interior of salon.

Fig. 2.7.4 "Burevestnik-24" aerodynamic ground-effect craft.

vessel floats on water as the regular boat, on 120 km/h the vessel begins to glide, reducing contact with water. And at a speed of 240 km/h rises over water thanks to the effect of wing-in-ground.

Ekranoplan WIG craft of CYG-11 developed by the Chinese firm Hainan Yingge Wing together with Russian engineers ("Ivolga" (Oriole) EK-12P project), can transport 10 passengers and two flight crews (Fig. 2.7.5) [58]. Configuration is based on the scheme of a catamaran and sea-plane on the airbag. Length of the ekranoplan is 13 m, wing span of 15.6 m,

Fig. 2.7.5 Patrol ekranoplan CYG-11.

speed of flight is 175 km/h, flight altitude of 3 m. The device is operated in the Chinese island province of Hainan in tourist, rescue areas, especially for sightseeing tours at sea.

Ekranoplan AirFish 8 developed by Wigetworks Private Limited (Wigetworks), was registered in Singapore in 2004 (Fig. 2.7.6) [59]. AirFish 8 is 8-10-seater ekranoplan with wing span of 15.0 m, length of 17.2 m, the speed of 180 km/h, flight altitude in the screen mode is up to 1 m. It is intended for management of two crew members and transportation of eight passengers for tourism and development of coastal infrastructure.

The Airfoil Development GmbH company (AFD) was founded in 1997 for development and production of ekranoplan [60]. Ekranoplan Hoverwing 20 is conceptual model with parameters: the maximum number of passengers is 23 persons, the maximal speed is 190 km/h, flying range of 900 km, length is 11.5 m, width is 11 m. In Fig. 2.7.7 the conceptual model of an ekranoplan Hoverwing 20 and the interior of passenger salon are presented.

Ekranoplan Flarecraft was developed by Flarecraft firm of the same name in 1997, the speed of flight of 150 km/h, flight altitude in the mode the screen up to 0.3 m over water, three persons containment (Fig. 2.7.8) is manufactured of composite materials [61].

Ekranoplan ARON-7 is developed by the South Korean company Aron Flying Ship Ltd. for rescue and military purposes (Fig. 2.7.9) [62]. The

Fig. 2.7.6 Ekranoplan AirFish 8.

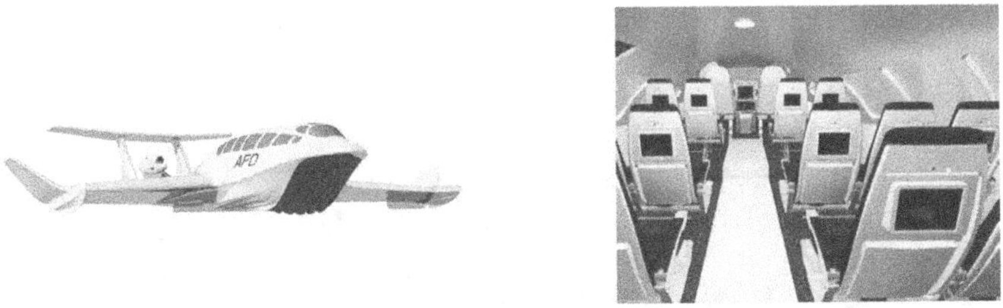

Fig. 2.7.7 Conceptual model of an ekranoplan Hoverwing 20 and interior of the salon.

Fig. 2.7.8 Ekranoplan Flarecraft.

ekranoplan can be located near coastal bases and quickly be applied to carrying out sea patrol. Ekranoplan (M50 version) has wing span of 12 m, length of 10 m, crew of 2 people, flight speed up to 200 km/h, flying range of 800 km.

In the presented review of ekranoplans generally small-size vessels of various producers who use the different modes of operation, both sea boats, and planes of amphibians were described. Let's consider also the

Fig. 2.7.9 Patrol ekranoplan ARON-7.

review of the modern publications devoted to research and development of ekranoplans.

2.7.3 Review of Publications

In the work [63] criteria and methods of synthesis of control systems for a heavy ekranoplan are described. The results received in the different countries and the prospects of creation of ekranoplan with autopilots are presented. Control algorithms have essential distinction at projection of small and larger ekranoplans. Ekranoplans, after the solution of a number of problems of the theory and practice of development, can apply for the essential sector of the market of high-speed vehicles.

The work [64] is devoted to digital modeling of aerodynamic characteristics of WIG craft with the engine (Power augmented ram PAR). The airflow blown from the engine accelerates a stream around the upper surface and strengthens it near a trailing edge of a wing. The simplified half of the model is applied for modeling thanks to symmetry.

In the work [65] aerodynamic characteristics of the WIG craft model of the ship are experimentally investigated. In a wind tunnel different configurations of the composite wing are checked: with trailer plates and without them. The wing-in-ground effect and existence of trailer plates increase coefficient of raising of a wing with a low flight altitude. Use of the WIG crafts is ambitious technology which will help to reduce costs of routine shipping in the future.

The work purpose [66] is the review of research and development of WIG craft technology of the ships. History and prospects of WIG craft technology, concrete vehicles, projects and ranges of application are considered.

Special attention is paid to aerodynamics of wing-in-ground effect, its mathematical model operation and stability of longitudinal motion. The questions bound to rules of classification, safety and certification are considered in more detail.

The article [67] is devoted to projection of a four-seater ekranoplan, and several options of the vessel are considered. As a result of the review, the inverse delta visible configuration which meets the requirements of design in the best way was chosen. The ship model was created, a test inspection of an aerodynamics of a design was carried out. As the key parameter of effectiveness of flight the criterion of decrease of length of a strip of takeoff is set. In the work [68] are investigated aerodynamic characteristics and also dynamics and stability of longitudinal flight of the WIG ship for improvement of the vessel's design.

The book [55] is devoted to the review of WIG technologies of effects and the historical review of theories and also approaches on development of WIG craft of the ships. In the book the following questions are considered: concepts of the WIG ships; stability, aerodynamic characteristics, power, quality and maneuverability and also the pilot studies; materials and designs, power supply, choice of engine installations; designs of the WIG ships, from the plane to design of the sea vessel, specifics of design of the concept of the WIG ships. Also the prospects of development of developments of the WIG ships are discussed, the principle of WIG can be applied to broad speed range and environmental conditions, that will lead to the fact that in the world of aircrafts both gliders and jet airliners will be used.

2.7.4 Concept of an Ekranoplan of "Water Strider"

For development of the idea of the concept the analysis of a visual number of inhabitants of water coastal water areas was carried out. The water strider, an insect with an original way of movement on water, was as a result chosen. Water striders (in Latin - *Gerridae*) (water strider) – falls into a family of semi-coleopterous insects from a suborder of bugs (Heteroptera) (Fig. 2.7.10) [69].

There are about 700 types of water striders. They live on a water surface and are considered to be the most widespread species in Europe. A routine pond water strider (*Gerris lacustris*) can fly by means of wings between reservoirs. The body and tips of the legs of a water strider are covered with rigid nonwettable hairs thanks to what they are adapted for, sliding on water. More short forward legs are used to hold a catch; the average pair of legs is used as rowing, and the back pair of legs control a water strider movement.

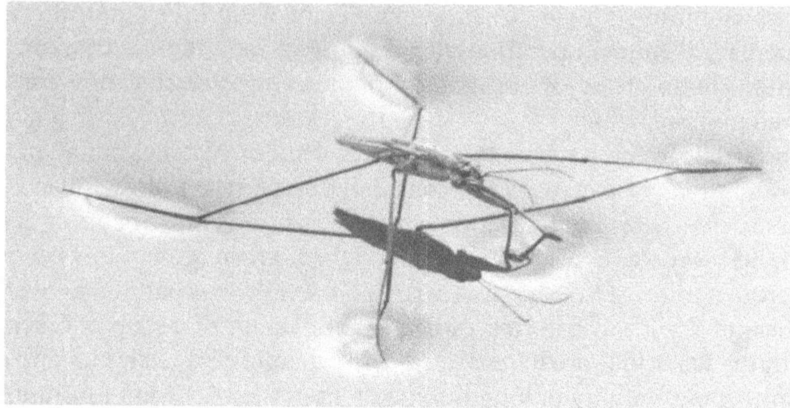

Fig. 2.7.10 Water strider.

In addition to good eyesight, a water strider also transfers and obtains information through fluctuations of the water surface. Such interaction is used also by males when searching for a female for pairing. Tiny hairs on the legs provide both a hydrophobic surface and a larger surface area to distribute the weight over water. As a result, water striders often move on a water surface with a speed about a meter per second.

Creation of new objects or products usually begins with development of the concept, creation of a prototype of future product [27, 37]. In a time of computer technologies this task becomes for a designer rather attractive, tempting for the imagination. At the beginning, the sketch, a portrayal of a future model, its composition, stylistic decision [38, 53] is created [39]. Then on the basis of the analysis of a visual number of natural forms the corresponding concept of future product is chosen.

It would be desirable to note that on this methodology in work [52] the original concept of the "Lapwing" amphibian was presented by the authors, and in the work [70] the concept of configuration of an interior of passenger inside of the amphibian, from the sketch to the ready three-dimensional model was described. In the course of working on vessel contours sketches natural forms were used, tracings of a contour of a water strider. Visual searching of a graphic image of an ekranoplan is given in Fig. 2.7.11.

The aircraft fuselage and furthermore the flying boat at the same time has to meet the requirements of aerodynamics and hydrodynamics. Therefore, the designer has to consider these features. The result of creative search work in creation of the space-spatial decision is presented in Fig. 2.7.12.

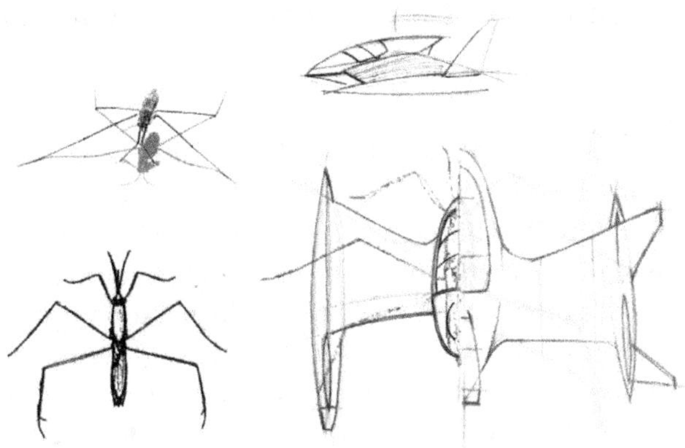

Fig. 2.7.11 Visual searching of a graphic image of the "Water Strider" ekranoplan.

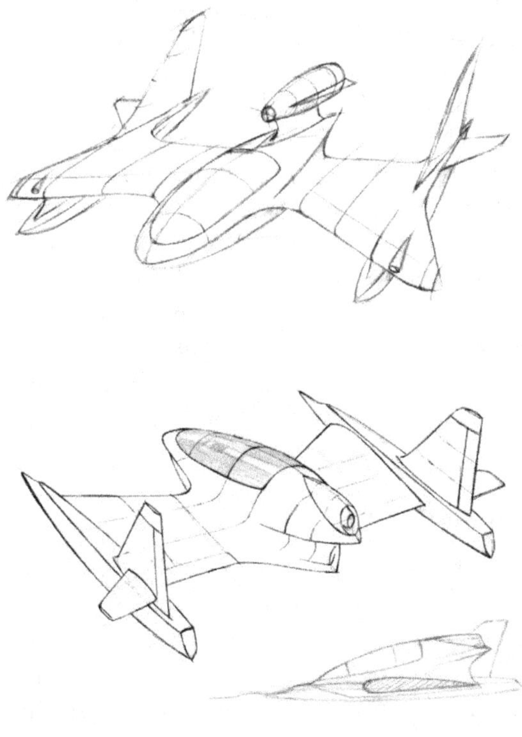

Fig. 2.7.12 Sketches of contours of an ekranoplan of "Water Strider".

2.7.5 Configuration of the Concept of an Ekranoplan

Concepts of an ekranoplan of "Water Strider" are the cornerstone of some technology solutions. The glider of an ekranoplan consists of the following constituents: cases, wing with flaps, skegs, rudders, stabilizers [19, 20]. The main designs of a glider are made of corrosion-resistant materials and alloys as the marine environment is rather aggressive for many metals. A power set of a glider (frames, stringers, ribs, walls) are made of composite material. The wing of an ekranoplan of an all-metal design has the trapezoid form of the inverse sweep with root flows (Fig. 2.7.13), consists of a center wing and two side consoles which are coming to an end of 4 section flap floats. Wing tips are served for increase in an efficient range of a wing and allow increasing lengthening of a wing, almost without changing at the same time its span. Vertical and horizontal stabilizers, rudders are located on side floats. For maneuvering the rudder in the tail of the center wing is provided in the mode of a gliding. In Fig. 2.7.13 projective types of an ekranoplan with overall dimensions are presented.

The sizes of a housing have to consider ergonomic requirements of future interior and task of accommodation of passengers [41, 42]. The wing span is 8.8 m, length of the plane is 8.3 m, height is 2.7 m. The crew consists of one person, in a passenger cabin are four passengers, and in the tail of the fuselage the small luggage compartment can be accommodated. The distinctiveness of an ekranoplan is the existence of panoramic glazing, which is made of athermal glass that allows increasing the review, both for passengers, and for the pilot. The engine installation is realized by two turbojets, which are located on poles closer to a tail part of the fuselage. Proceeding from the sizes of the wing chord dimensions, it is possible to note that flight altitude in the mode of the screen will be up to 0.5 m over water. Ekranoplan can be applied to transportation of passengers, patrol, environmental control of water areas, rendering urgent medical care, providing a wrecking, rest and tourism.

2.7.6 Stages of Modeling

After development of the drawing we pass to stage-by-stage modeling of an ekranoplan. For creation of three-dimensional computer model there are many various graphic systems. In our case we will use the most comprehensive graphic system of three-dimensional modeling 3ds Max, which has ample opportunities on modeling and photorealistic rendering [71]. The created model for further development can be exported to graphic CAD, the computer-aided engineering system.

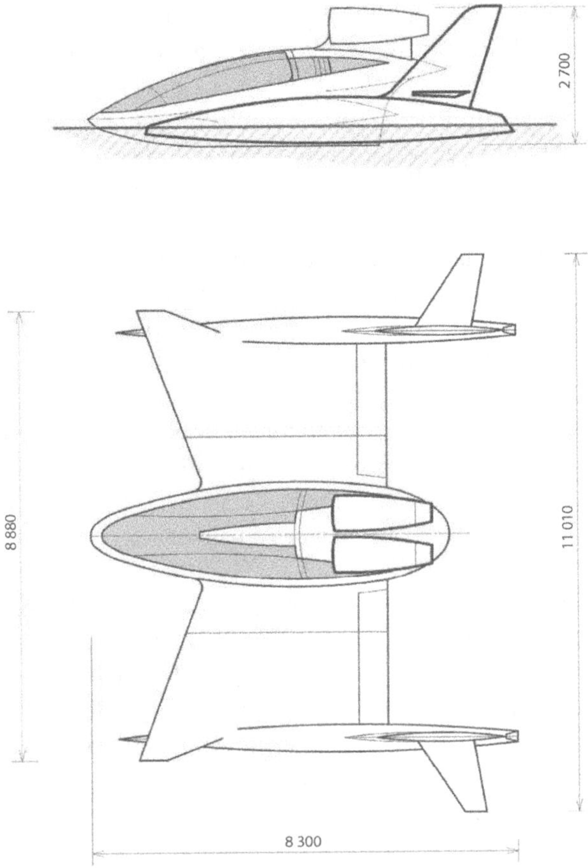

Fig. 2.7.13 Projective types of an ekranoplan with overall dimensions.

For creation of a three-dimensional model of an ekranoplan it is possible to use several methods, but process begins with the available schematical projections of the modelled object. On three mutually perpendicular planes projective drawings are placed, and creation of an object is carried out by rather visual teaching method of a polygonal extrusion. As the tentative polygon, the three-dimensional primitive Plane from which the fuselage and constituents of the ship is created is used. According to a method of a serial polygonal extrusion it is necessary to keep the constant number of polygons along all fuselage, for an exception of problems with geometry and a possibility of further completion of model. Therefore, we create the minimum quantity of polygons initially; later, at a shortcoming, they can be added by a section method the operation Slice polygon. The process of an extrusion is strictly controlled in compliance to drawings of contours of the fuselage on all three planes.

In Fig. 2.7.14 the initial preparation of model of the fuselage is presented; other parts of the model will also be squeezed out from this preparation. Then we pass to creation of the following important part of the future ship, the wing.

The wing of this plane has the composite profile as it carries out the bearing role for the plane in the mode of a gliding and it works as the screen, increasing lift force at the time of takeoff from the water surface. The same way, separately, the bearing pole of the engine and the engine (Fig. 2.7.15) is carried out.

The design of an ekranoplan includes, besides the fuselage and the bearing wing, side floats. Vertical stabilizers which perform the function of vertical plumage of the plane are integrated into cases of floats. Modeling of a housing of a float is carried out by an extrusion on a contour of the general view of a product (Fig. 2.7.16).

At the following stage we carry out operational development of geometry of objects, we smooth all constituents of future model by NURMS subdivision method within the Editable poly frames; the result of application of this function considerably improves the appearance of an initial angular form of preparation [54]. Then the stage of assembly of the plane from constituents follows. The pole (rack) for the engine and a motor case join the fuselage with a wing and tail plumage. In addition, on a wing and tail plumage marker lights settle down. Specification of the

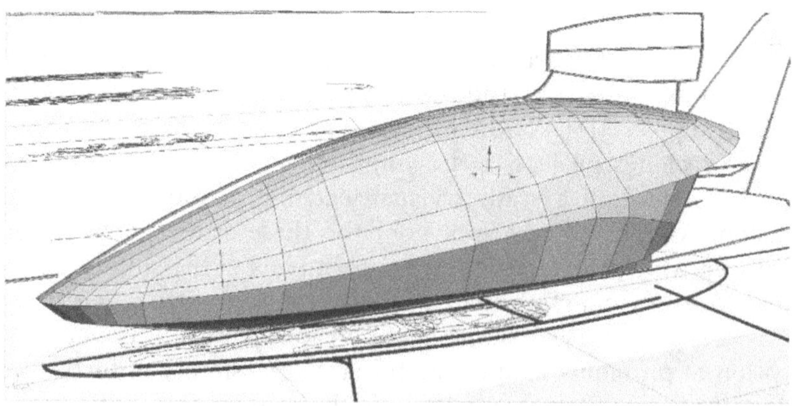

Fig. 2.7.14 Preparation of future fuselage.

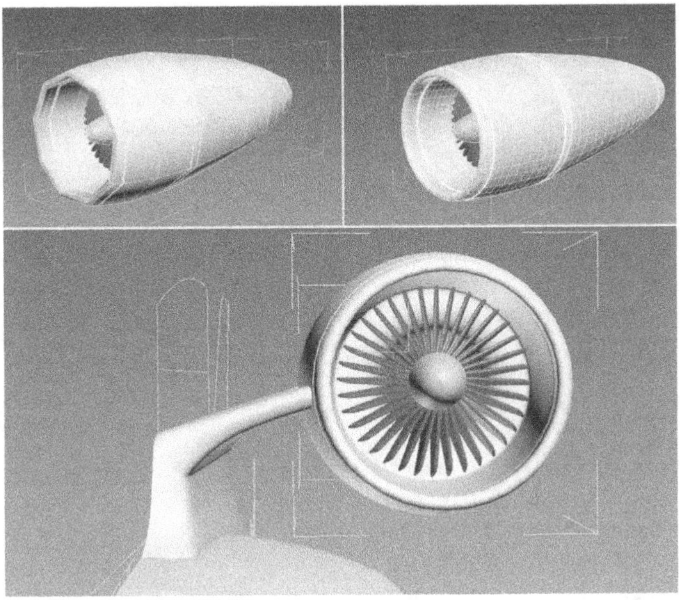

Fig. 2.7.15 Modeling of engine.

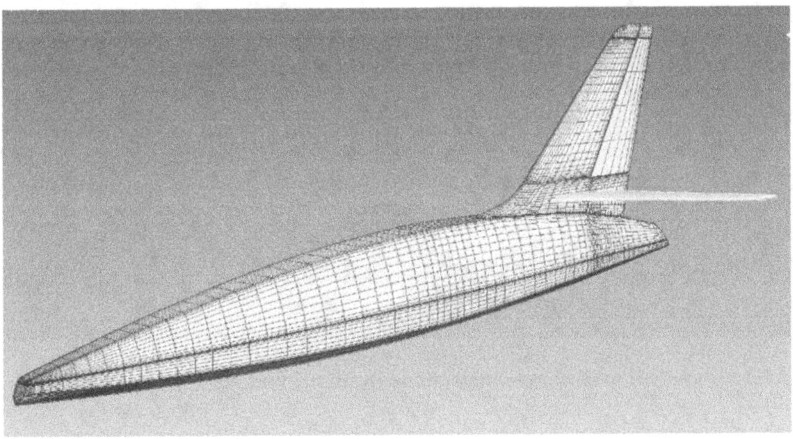

Fig. 2.7.16 The finished model of a side float.

fuselage means modeling of wind glazing and side windows. The wing along with a wheel of management and the horizontal stabilizer is also exposed to more careful detailing. Thanks to symmetry of a design of the ship, the process of modeling is conducted with one half of an ekranoplan (Fig. 2.7.17).

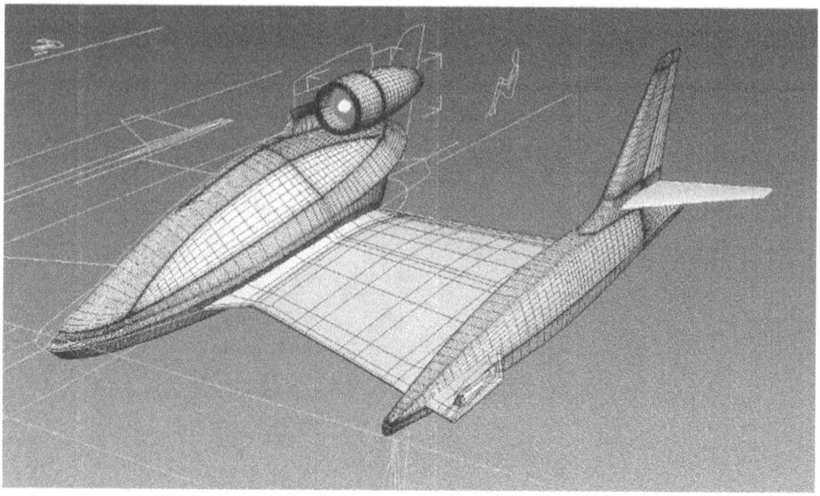

Fig. 2.7.17 Smoothed model half of a housing.

Fig. 2.7.18 Assembling all components of the ekranoplan model.

To create a complete conceived ekranoplan, we supplement the scene with the second half relative to the longitudinal axis. The result of the final model of the ekranoplan is shown in Fig. 2.7.18.

2.7.7 Shading and Rendering of Model

The following stage concludes turning of model for the subsequent rendering. To assign materials for design parts of a housing, we begin the process

Fig. 2.7.19 Rendering of an ekranoplan "Water Strider" in the gliding mode.

of selection of necessary elements at the level of editing polygons, in turn appropriating earlier adjusted materials to model constituents. In our case for receiving photorealistic rendering the choice of material and control of its main properties, taking into account irradiating, is carried out by means of the external module of rendering for 3ds by Max V-Ray from the ChaosGroup developer. Therefore, for shading of objects the library of the materials VRayMtl will be used.

Details of the process of creation of materials and installation of irradiating will be considered in the section of rendering of an interior of an ekranoplan. Rendering of a scene with ekranoplan "Water Strider" in the gliding mode is presented in Fig. 2.7.19. In this drawing the photorealistic model of an ekranoplan is rather visually shown during takeoff over the water surface; it is also possible to analyze configuration of external contours of the ship. For development of the complete complex of the model of an ekranoplan further we will pass to the development of an interior of the passenger salon.

2.7.8 Development of an Interior and Passenger Chair

In spite of the fact that an ekranoplan is intended for flight over average distances, questions of ergonomics and comfort of an interior of the passenger salon remain important [26]. Also it is necessary to consider the fire danger of materials of finishing at projection of an interior of a cabin.

For creation of a realistic and reliable model of the fuselage we will use the drawing with cuts, sections of future case which is presented in Fig. 2.7.20.

Modeling of separate elements of the salon is executed also by the method of a polygonal extrusion. This configuration of salon with one crew member and four passengers allowed the placing of individual control panels of blowing, adjustment of temperature and the block for connection of multimedia of devices through a garniture. The passenger chairs deserve special attention; they are executed taking into account ergonomics and normative requirements, and their convenience sometimes is the major factor in the choice of the vehicle. Development of a chair begins with sketches, the ergonomic scheme of a chair (Fig. 2.7.21) with anthropometric parameters [50] further is developed [47]. Serial stages of model operation of a chair are presented by the method of a polygonal extrusion in Fig. 2.7.22.

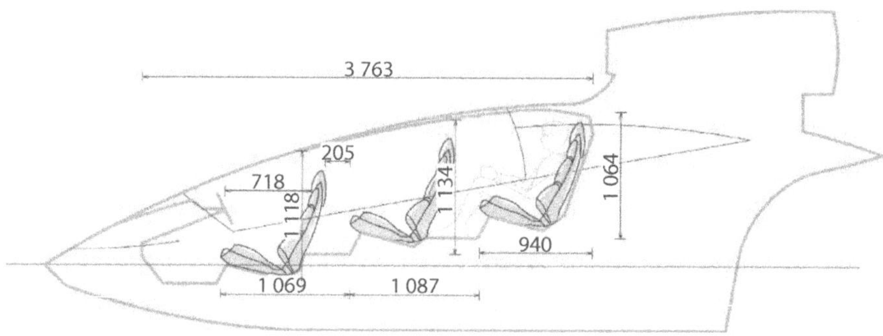

Fig. 2.7.20 The drawing of salon of an ekranoplan in a section.

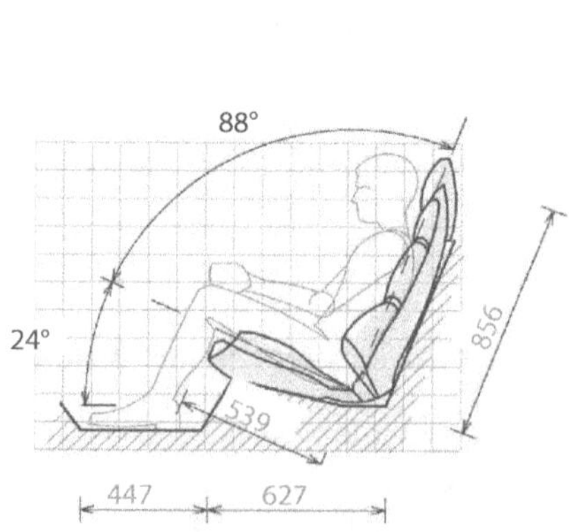

Fig. 2.7.21 Ergonomic scheme of a chair.

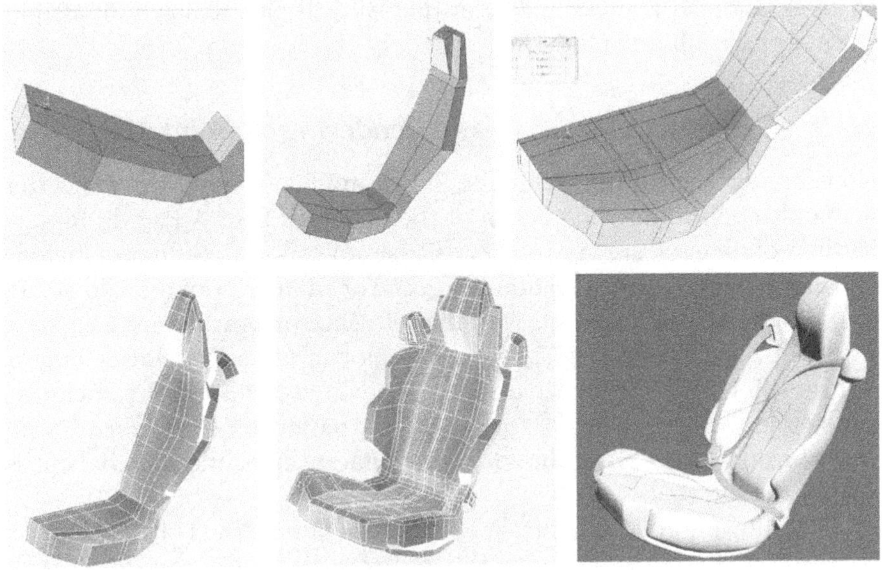

Fig. 2.7.22 Stage-by-stage creation of model of a chair.

Further specification of a configuration and operational development of the model of a chair is made, and at a final stage we use the Smooth modifier for smoothing of polygons. According to the scheme of configuration of the salon it is necessary to copy and distribute chairs. The convenient Array team is used for this purpose. The window of this order allows to set values of shift of the subsequent copies relative to each other, on all three planes simultaneously. In Fig. 2.7.23 the closing configuration of the salon

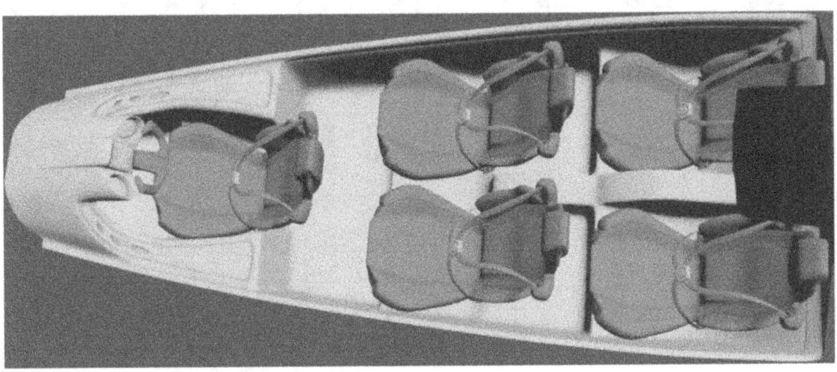

Fig. 2.7.23 Configuration of salon of an ekranoplan.

of an ekranoplan is presented; after that we will pass to stages of assignment of materials and rendering.

2.7.9 Creation of Materials and Rendering of an Interior

Considering that for rendering the V-Ray module is used, this visualizer demands application and control of the materials from VRayMtl library. Settings of material are in a roll of Basic parameters of the editor of materials (Fig. 2.7.24). During creation of a scene of an interior of the salon of an ekranoplan, several materials with the reference parameters were applied. In the option Diffuse (Fig. 2.7.24), the color of the surface dispelling or diffuse color which is reflected is set. The editor of materials can choose simple monochromatic color for the Diffuse parameter or use the textural card. In our case we use the textures imitating finishing materials; it is leather, velour, rubber.

The following roll for control of properties of material is Reflection. As is already clear from the name, this roll sets the reflecting properties of material. All surfaces in the world around have the reflecting properties. If to choose purely black color, then it deprives the surface of material of reflections; white, on the contrary, does completely reflecting. All intermediate values of gray color influence reflection force. In a window it is possible to set values from 0 to 255, and it means to receive material with the reflecting ability of 50%, it is necessary to establish value 128.

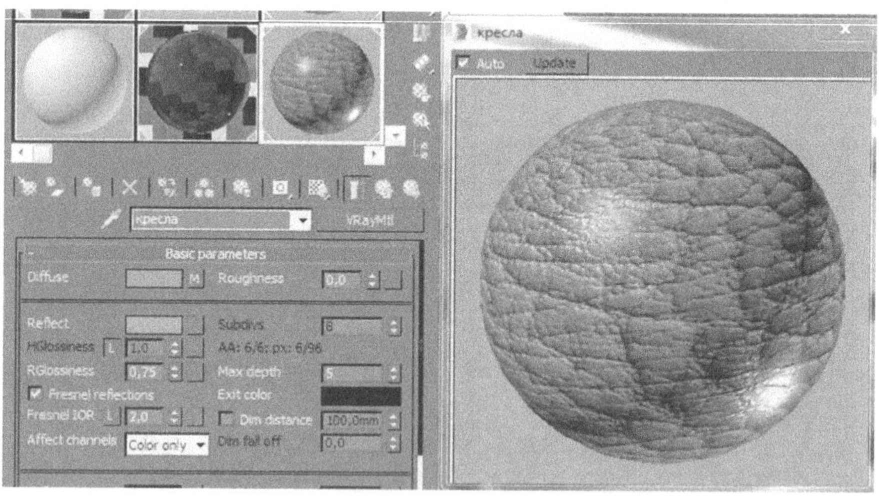

Fig. 2.7.24 Roll of basic parameters of the Editor of the materials VRayMtl.

Except control of above-mentioned parameters, we adjust the invoice of material of chairs; for the card of a relief Bump we use the same texture, as in the Diffuse option, but only the image in gray tones (Fig. 2.7.25).

Other materials are created the same way, except materials of fabric upholsteries, reflection at such materials is practically absent. One of such materials is the valor's upholstery of panels between chairs. Use of materials to model is made at the level of the respective grounds. In the course of model operation, polygons for the corresponding materials (Fig. 2.7.26)

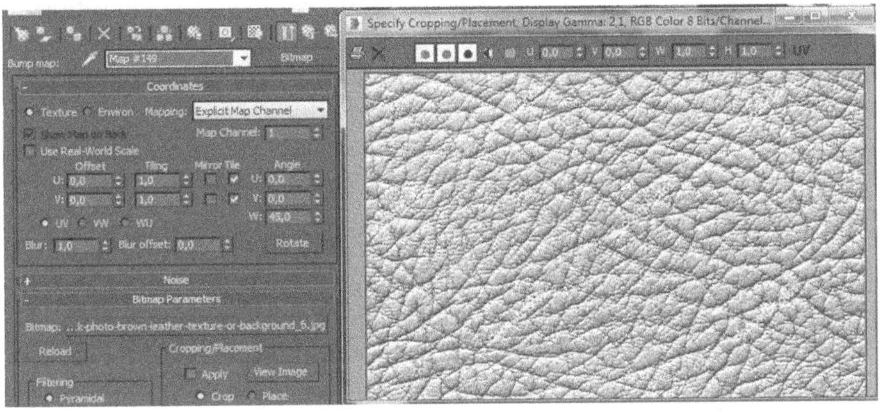

Fig. 2.7.25 Control of parameter of a relief Bump of the material VRayMtl.

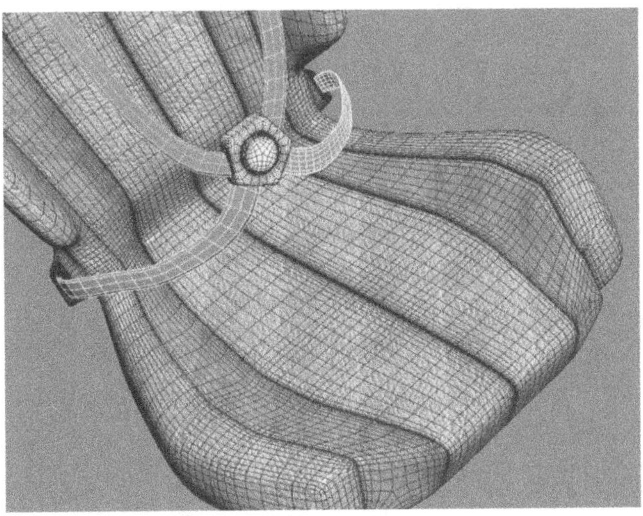

Fig. 2.7.26 Use of materials to chair model.

were provided in advance. It is possible to use the modifier taking into account UVW Map geometry to correct display of texture of upholstery. This modifier is used in the Box mode that allows stretching or squeezing texture in all three planes.

For a terminating miscalculation of a scene and receiving photorealistic rendering, the module of rendering V-Ray gives the user many opportunities. This process is quite variable: various materials, various geometry of models and an environment in a scene, influence quality of final rendering and time of a miscalculation of process. For high-quality display of model to rendering, it is necessary to establish light sources. The algorithm of global illumination V-Ray gives access to a number of means of lighting. Global illumination of GI Environment allows to adjust a luminescence of a surrounding medium. In this scene of a luminescence of the environment it is not enough; therefore for better lighting we will place light sources of VrayLight of the Plane type, light from this source spreads in the direction of an arrow (Fig. 2.7.27).

It is possible to configure key parameters of a light source for a specific objective: Intensity – intensity of a source; Temperature – its color temperature in kelvins; Cast shadows – sets the mode of a deletion of shadows. After control of necessary parameters the scene with an interior of salon of an ekranoplan, which is presented in Fig. 2.7.28 and Fig. 2.7.29, was rendered. As a result of its development it is possible to note that the submitted

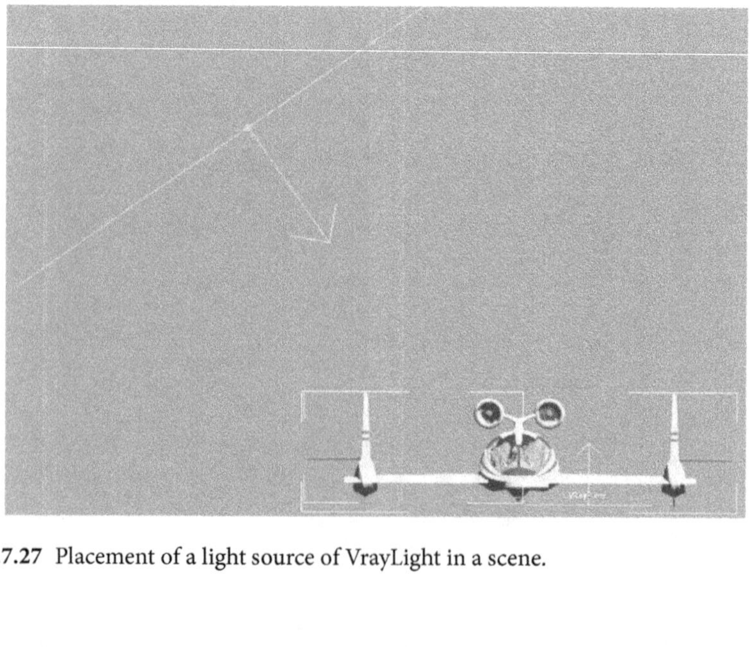

Fig. 2.7.27 Placement of a light source of VrayLight in a scene.

Fig. 2.7.28 Rendering of salon of an ekranoplan "Water Strider" with the pilot's cabin.

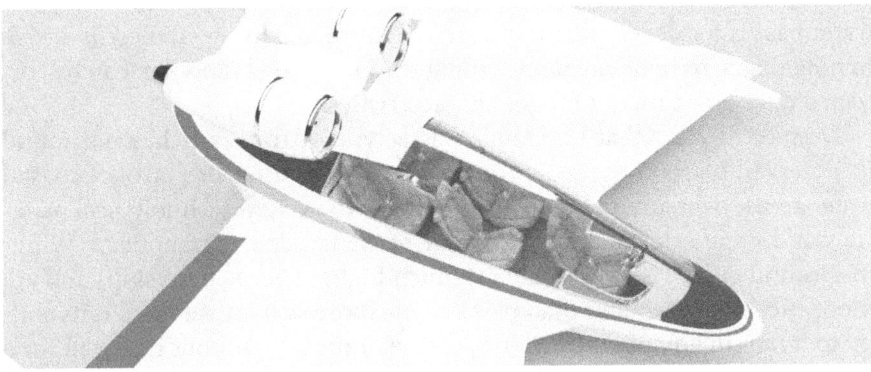

Fig. 2.7.29 Rendering of salon of an ekranoplan "Water Strider".

concept of configuration of salon and its model operation allows the creation of quite comfortable modification of the salon of an ekranoplan.

As infrastructure of sea basing, in Fig. 2.7.30 the scene of rendering with the module of landing of passengers for an ekranoplan of "Water Strider" for development of tourism and voyages is presented.

2.7.10 Conclusion

In conclusion it is possible to note that in this work the concept of the water vehicle of an ekranoplan of "Water Strider" was presented, beginning from the initial idea, sketches before development of terminating three-dimensional model with projection of an interior of salon [72]. This concept is ready for realization in geographical regions with branched

Fig. 2.7.30 Rendering of a scene with the landing module of an ekranoplan "Water Strider".

water basins for development of regional tourism. However, there is a wish to note that aero hydrodynamic configuration of an ekranoplan can be corrected after assessment of flight characteristics.

Despite the fact that ekranoplan as a type of transport has not found its deserved place in the sector of the tourism industry or patrol of coastal water areas so far, undoubtedly its advantages with new technological materials and engine installations will still be demanded. Ekranoplan (Wing-in-Ground craft) move on water at much higher speeds than ships and are more effective over water than planes. One more of its advantages is its ability to fly up in any place from the surface of the sea without the need for a runway. Thanks to the excellent and unique properties uniting possibilities of the boat and the plane, ekranoplans have the prospect of rapid and efficient growth among the new generation of sea transport systems.

2.8 Design of Multifunctional Hydrofoil "Afalina"

2.8.1 Introduction

The leading coastal countries of the world need to create and build new vehicles for the development of their economies. Complex logistical tasks for many developed countries have been solved using various types of infrastructure. These tasks include air transportation, land rail links, highways and the country's vast water network. The task of delivering passengers and cargo in a short time over long distances, however, remains urgent.

This paper presents the process of creating a design concept for a multifunctional hydrofoil. The idea of this project is to create a universal multifunctional platform, on the basis of which various configurations can be applied as soon as possible, depending on the purpose of the transport vessel.

In the middle of the last century, vessels with dynamic principles of support were developed in Russia. For this purpose, the wing effect was used to move over the surface of the water; they were called hydrofoil vessels and ekranoplans. These vessels use the general principle of movement – raising the hull above the water with the help of wings, which makes it possible to reduce water resistance to the movement of the vessel and, thereby, significantly increase the speed of movement.

The hydrofoils are vessels that use the lifting forces that arise when water flows along the surface of the wing. A hydrofoil vessel is a type of high-speed vessel using the effect of a submerged hydrofoil (the Alekseev effect). The Alekseev underwater wing consists of two main horizontal bearing planes—one in front and one in the rear [73, 116]. The "wings" are submerged in such a way that the ship's hull is above the water while it is moving, which allows the ship to go much faster, since water is more than a thousand times denser than air [117]. Initially, hydrofoil vessels were attached to small hulls and had mostly experimental interest. Today, several large hydrofoil passenger ferries are operating around the world, which are economically sound and are used mainly in the coastal countries of Europe, in Canada and in Russia. These modern vessels remain effective, as they have good seaworthy characteristics and therefore are used for commercial cargo transportation.

2.8.2 Research Overview

Consider some modern publications in this field and conceptual models of ships with underwater wings. In the work [74] mathematical models of the movement and stability of hovercraft and hydrofoils are considered. A mathematical model was developed in [75] to optimize vessel stability factors to prevent overturning and increase the maneuverability of a hydrofoil vessel at high speeds. For these purposes, the proposed model offers the use of mobile wing technology.

The work [76] is devoted to the optimization of the basic elements of high-speed catamarans with hydrofoils at the design stage; their advantages and disadvantages are also analyzed. The aim of the work [77] is to improve the performance of the displacement catamaran on hydrofoil with a wind power plant. The hydrodynamic model is based on the Reynolds-Navier-Stokes equations (RANS). The calculation results showed energy savings when using the technology of underwater wings.

In [78], the hydrodynamic characteristics of hydrofoil vessels are investigated using the discrete vortex method. A method of mathematical modeling of the dynamics of the movement of a hydrofoil for calculating resistance, landing, stability, controllability and strength is proposed. In [118], the turbulent flow around hydrofoil vessels is investigated, and the averaged Navier-Stokes method is used for modeling. The calculation results are compared with experimental data.

The paper [79] presents an analytical review of the historical experience of operating ships on hydrofoils, and describes their current state and prospects for the development of the fleet of cruise ships. Original sketches and models of hydrofoil ships Alekseeva are presented on the site [80].

Consider the modern models of these vessels, which are developed by the Nizhny Novgorod Central Design Bureau for hydrofoil vessels named after R.Ye. Alekseev (CDB for Hydrofoil). These ships include the marine passenger hydrofoil ship "Comet 120M", created in 2017 [73] (Fig. 2.8.1, left). It is intended for high-speed passenger transportation during the day in salons with aircraft seats. This is a single-deck vessel, with a diesel-reducer power plant, the length of the overall m 35.2 m, the number of passengers 120 people. It provides the movement of the vessel in the wing mode with a wave height of up to 2.0 m and a wind of up to 4 points on the Beaufort scale.

It is also possible to note the small-seating river passenger hydrofoil "Valdai 45R", which was created in 2017 [73] (Fig. 2.8.1, right), overall length is 21.3 m, the number of passengers is 45 people. The movement of the vessel in the wing mode is provided at a wave height of up to 0.7 m and a wind of up to 2 points on the Beaufort scale at a speed of 45 km/h.

The project of a large passenger marine hydrofoil "Cyclone 250M" was developed by the Alekseev Central Design Bureau [73]. The vessel will have a length of 42 m, and carry up to 320 people on board. The power plant of two engines will provide speeds of up to 55 knots (101 km/h) and

Fig. 2.8.1 Ships on the hydrofoil "Comet 120M" "Valdai 45R".

a cruising range of 700 miles (1,300 km). Promising areas of operation of the "Cyclone 250M" are: Southeast Asia, the Russian Far East, the Baltic and Black Seas, and the Caspian Sea. With a wind of up to 4 points, a wave height of up to 3.5 m ensured the safe navigation of the vessel in a displacement mode.

The article [81] presents an interdisciplinary design method for a conceptual model of a large-sized high-speed hydrofoil. This method is used to determine the maximum lift force at high speeds. The paper [82] describes the economical river high-speed vessels of the new generation "Strela M1", developed by the Krylov Research Center, the number of passengers is 60 people, the speed is up to 50 km/h for use in sparsely populated areas.

The work [83] presents the results of full-scale tests of the model of a hydrofoil ship. The model was equipped with two lifting and stabilizing front wings. The purpose of the work [84] was the development and creation of a prototype hydrofoil ship compatible with a conventional sailing boat. Using mathematical modeling, the design of the vessel was refined and a prototype was built.

The work [85] is devoted to building a hydrofoil vessel, which combines the simplicity of classical designs and modern innovative developments (Fig. 2.8.2, left). The vessel is equipped with aluminum front hydrofoils of the V-shaped configuration and aft T-shaped configuration. In the work [86] a light motor boat with underwater wings on electric batteries is presented (Fig. 2.8.2, right). The absence of emissions allows it to be used in places where water transport on gas is prohibited. The hull of the ship is made of composite materials; the C-shaped legs on hydrofoils are made of aluminum alloy. Inside, the ship has an ergonomic high-class cabin with all basic functions, the steering wheel is designed in the style of Formula 1 with a color touch screen, and passenger seats are designed for comfort and visibility during the cruise.

Fig. 2.8.2 Experimental layout of a hydrofoil vessel [85] and a light motorboat [86].

2.8.3 Development of the Concept

Let us proceed to the development of the design concept of the vessel. For this, on the basis of an associative array (Fig. 2.8.3), sketches of hull lines and forms as a whole were developed. In the process of the work, it is necessary to adhere to the requirements of ergonomics and functionality [41, 42], based on these requirements, some visual solutions were found of various configurations.

The results of the search for spatial solutions are presented in the sketches (Fig. 2.8.4).

As a result of the analysis of the vessel's exploitation environment, a dolphin was chosen as a biological prototype, and the project was named "Afalina". A bottle-nosed dolphin, an afalina or a large dolphin, is a species of dolphin, a cetacean group, that belongs to marine mammals [119]. The bottlenose dolphin usually does not exceed 3 m, the weight is 300 kg, their moderately developed "beak" is clearly limited from the convex fronto-nasal pillow, the body is dark-brown from above, light-colored from below.

The design concept is planned for coastal sea cruises; therefore the design length of the proposed vessel will be 44.3 m, width 13.7 m, the vessel is designed to carry 100-150 passengers, depending on the configuration of the cabin, cruising speed up to 100 km/h (Fig. 2.8.5). The vessel has entrances on the sides and aft, and it is equipped with hermetic doors and rescue rafts. According to the concept, the hull is a monoblock with surges at the stern. Floods form compartments for power plants, along with the left- and right-side tanks for fuel are assumed. The power plant is

Fig. 2.8.3 Associative series to create a design concept.

Fig. 2.8.4 Sketches of the prototype ship.

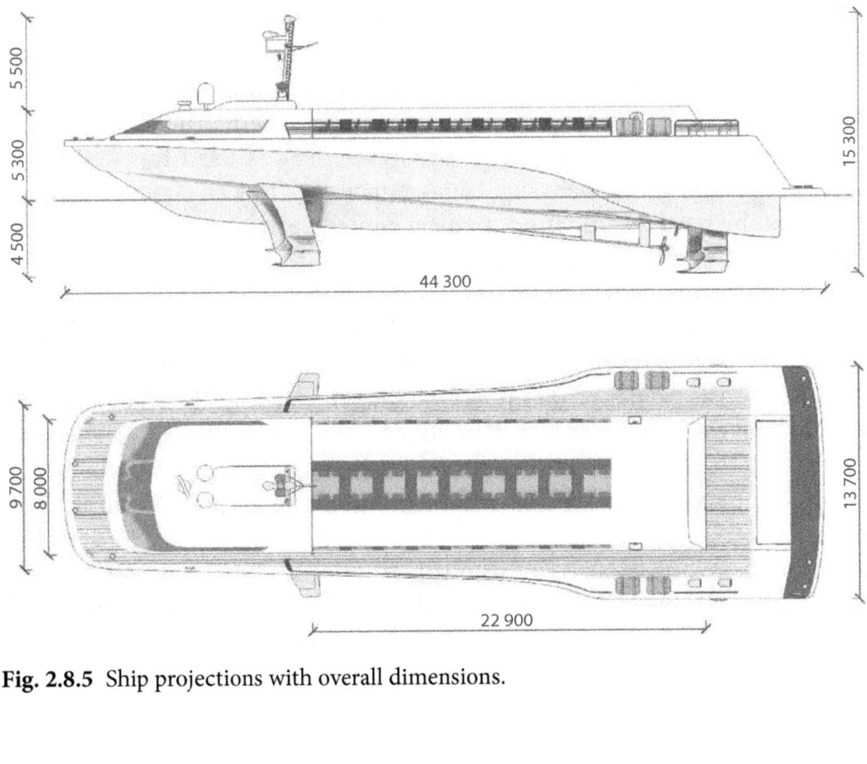

Fig. 2.8.5 Ship projections with overall dimensions.

represented by three motors – right, left and center. In the cargo configuration, the hull is a rectangular cabin and cargo hold. In the case of using the vessel as a cargo, the upper passenger module is removed.

2.8.4 Ship Modeling

Previously, the implementation of models in a real project took a lot of time and effort; today, in the digital age, digital assistants come to the aid of specialists in the form of software systems and 3D printing of the layout. Along with many 3D modeling programs, 3D Studio Max allows you to develop projects of almost any complexity [120]. This graphic system allows you to work with drawings made in other graphic packages, thereby giving the user plenty of room to work. To create a three-dimensional model, you can use several methods that require the construction of schematic projections of a simulated object [71]. The modeling process begins with the creation of three perpendicular planes, with images of projections placed on them.

Next, create a set of sections of the fuselage, which are placed along the axis of symmetry. A spatial grid is formed on the basis of these sections. The next step is to create a surface on the basis of the created grid (Fig. 2.8.6). The constructed grid in the process of finishing the model is slightly edited; all the component parts of the vessel are worked through in the same way, with subsequent adjustment by projections.

The unification of the modules allows you to change the destination of the vessel, the cargo-and-passenger version represents the vessel on the principle of a sea ferry, on which the lower deck is used for vehicles or other cargo, and the upper deck is passenger (Fig. 2.8.7). As intended, the vessel can be used as a floating medical center for the needs of rescue services, and it is also possible that the armed forces can use it to quickly transfer personnel and equipment.

For the rescue configuration, a mobile hospital module is installed in place of the passenger module; a cargo module is used for the cargo version (Fig. 2.8.8). The power structure is a U-shaped scheme, closed at the stern lifting ramp for loading vehicles on board.

It should be noted that the feature of this project is the combination of the classic single-hull fuselage scheme with elements of a trimaran. This allows you to increase the stability of the vessel at low speeds, besides this three-point layout is based on the tunnel effect, which is created by the front part of the hull with two floats. The use of modular layout allows you to increase the versatility of the vehicle and makes it economically advantageous to operate.

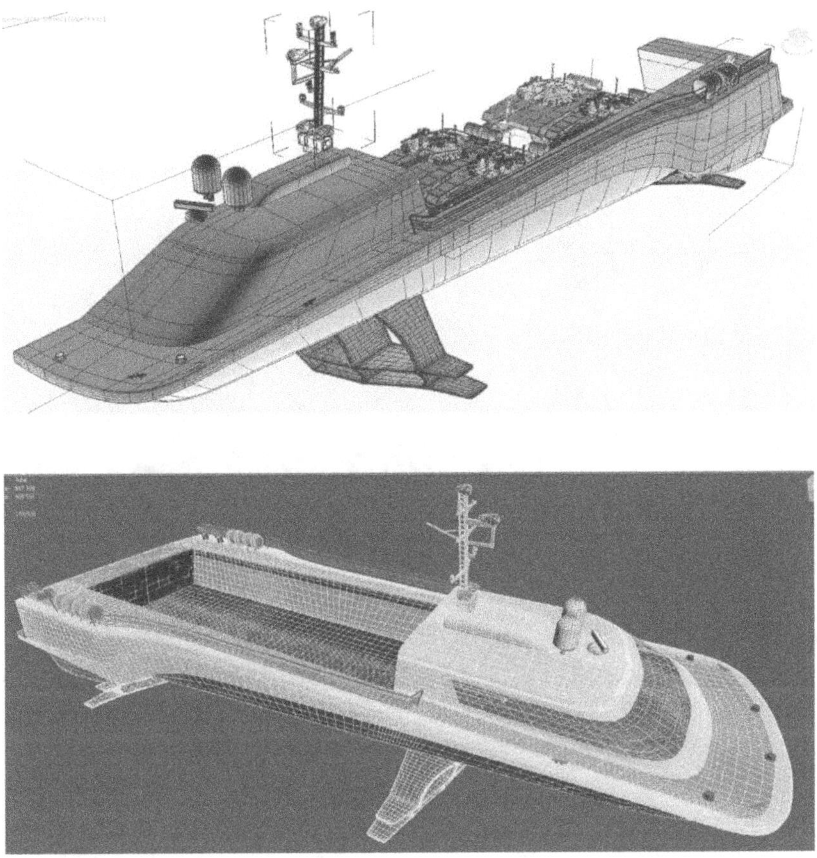

Fig. 2.8.6 The result of the creation of a three-dimensional model of the vessel.

2.8.5 Shading and Rendering of the Model

After assembling all the component parts of the model, it is necessary to set up and assign materials, and set the stage lighting for the final rendering of the prototype. To create photorealistic rendering of 3ds Max, an external rendering module V-Ray is used. This renderer requires the use and customization of its library of materials VRayMtl, which differ from the standard materials of the program itself. When creating the model, several materials with characteristic parameters were applied. Fig. 2.8.9 shows the process of creating a varnished coating of the material of the upper deck.

For rendering the scene and receiving photo-realistic images, the V-Ray renderer has many settings. This process is rather variable; different materials, different geometry of models and environments in the scene affect

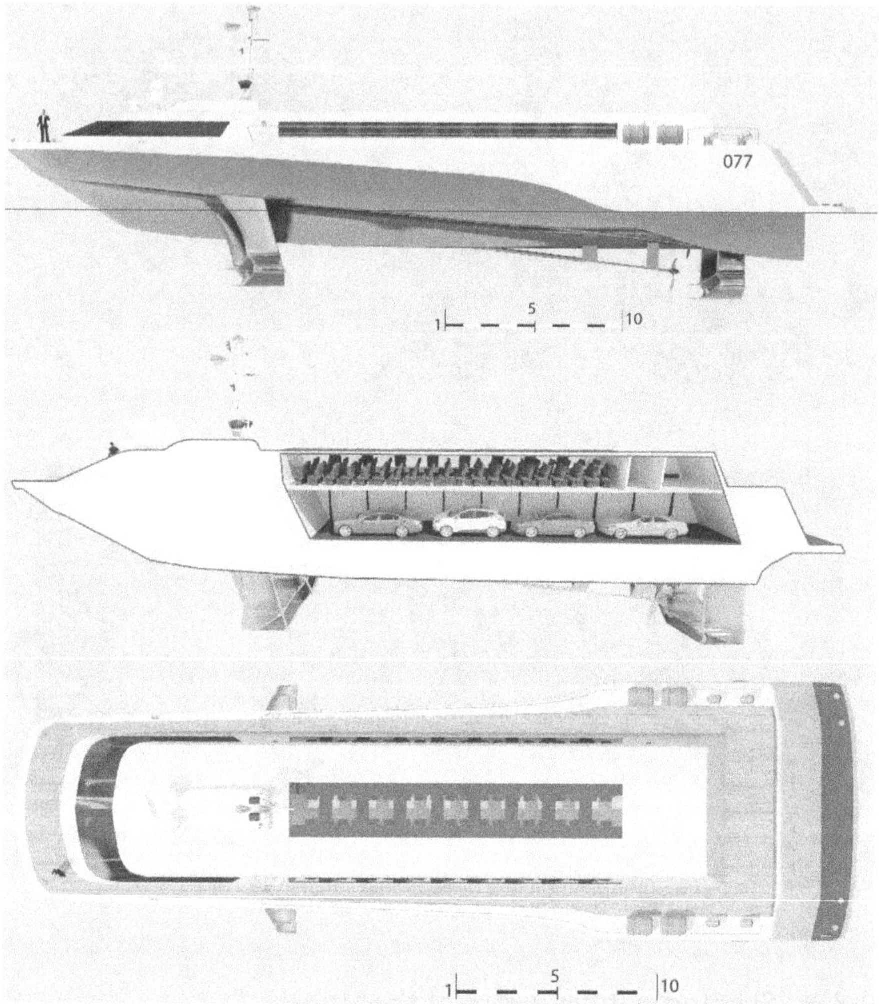

Fig. 2.8.7 Passenger/cargo layout.

the rendering time and the quality of rendering. In Figure 2.8.10, at the top are rendering of flat views behind and in front of the ship model; at the bottom of the figure is a three-dimensional view of the model showing the design features of the hydrofoil ship. The result of the final rendering of the ship model after the graphic processing in the operating environment is shown in Fig. 2.8.11.

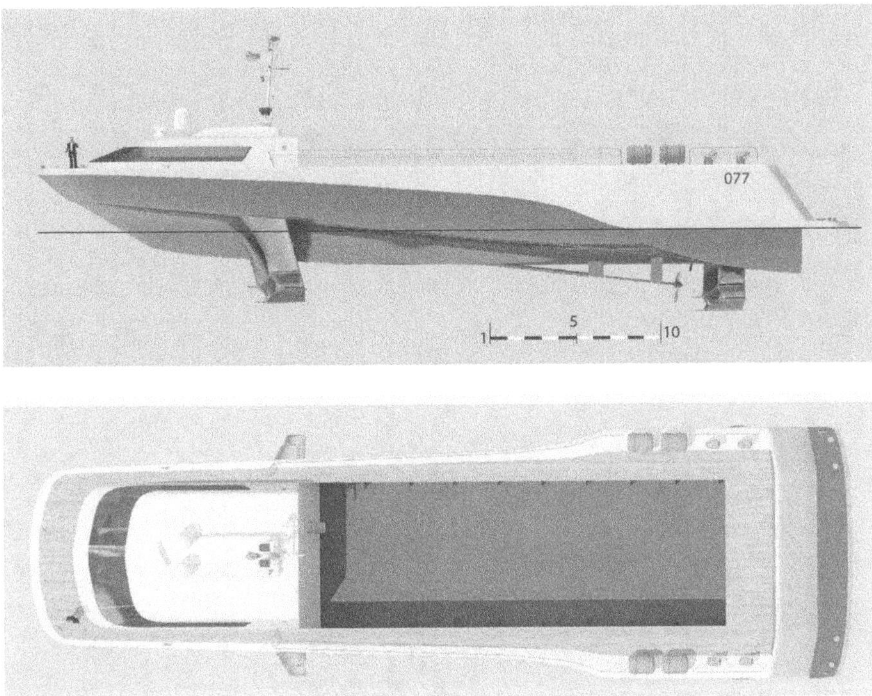

Fig. 2.8.8 Cargo layout.

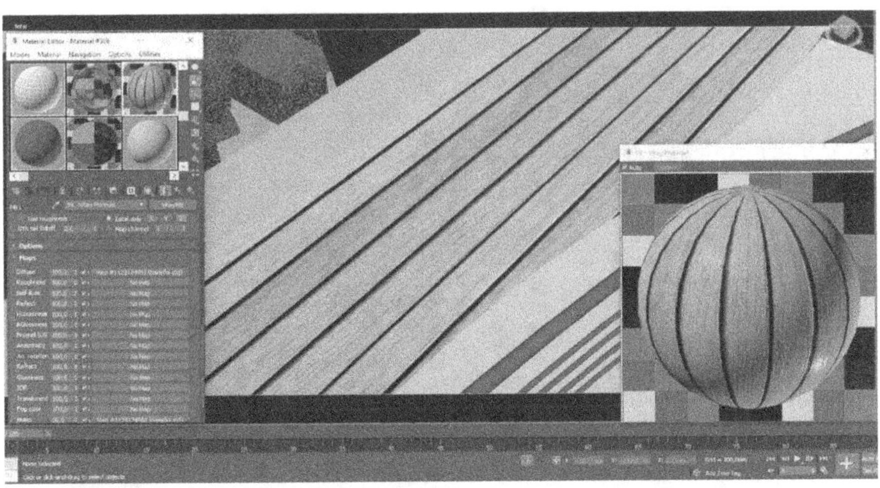

Fig. 2.8.9 Creating a VRayMtl material in the material editor.

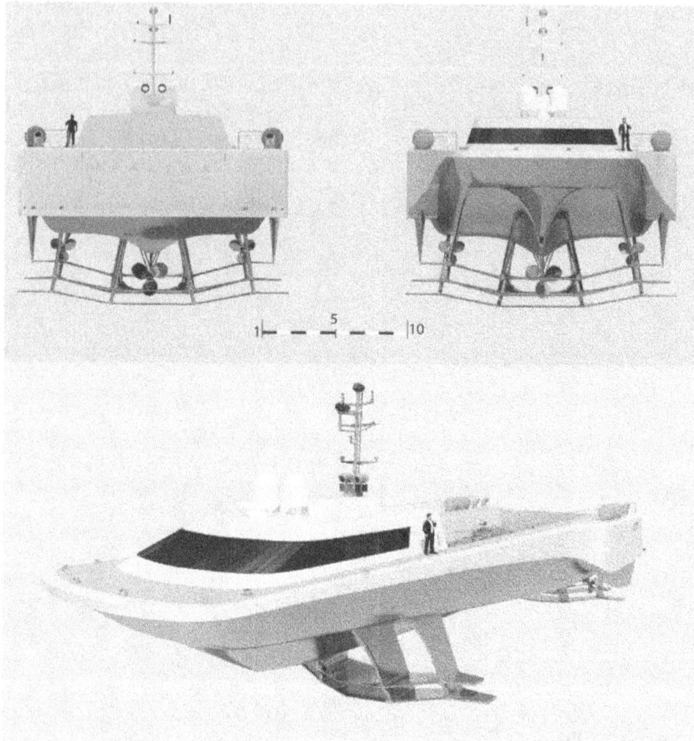

Fig. 2.8.10 Three-dimensional views of the model of the hydrofoil "Afalina".

Fig. 2.8.11 Photorealistic rendering of the hydrofoil model "Afalina" in the exploitation environment during a sea cruise.

2.8.6 Conclusion

The results presented in this paper on the development of a prototype hydrofoil are mainly theoretical. The paper presents the development of hydrofoil ship "Afalina", ranging from concept to the final rendering of the finished model. Similarly, an ekranoplan, a hydrofoil vessel, is designed for high-speed delivery of passengers or other cargo within the coastal waters. The advantages of these vessels include their excellent seaworthy characteristics, multifunctionality and profitability.

2.9 Autonomous Mobile Robotic System "Sesarma"

2.9.1 Introduction

At present the robotic systems in human life have actual meaning, more than ever. In many fields of our life robotic systems of different degrees of complexity are used as assistants. In the majority of cases robots come to the aid of man where access for people is impossible or is connected with defined life and health risks. Many developed countries conduct investigations on the development of mobile robots for different purposes. At present in the processing field meaningful results have been achieved connected with the creation of high-power onboard computers, compact executive mechanisms and intelligent sensors. The modern algorithms use high-speed information processing methods when controlling motion, both of single mobile robotic systems, and within a group.

In this paper we review the design and the carrying out of computer modeling of autonomous four-wheel mobile robotic system to solve different tasks. When solving tasks in the field of industrial design the preliminary elaboration and selection of design of the product body are rational. The developed mobile robotic system has a three-level control system and can operate within a group when moving under data incompleteness conditions, in the nondeterministic medium. Let us consider some review of the recent research in this field and the current analogues of the created robotic system.

2.9.2 Review of Publications

The paper [87] is dedicated to the analysis of the current state of global trends in the area of creation of the small-size robot support facilities of different assignment. The basic principles and development prospects of mobile mini-robots are highlighted. The paper [88] presents an overview of research in the field of autonomous mobile robot control, including the possibility of using the principles of interpretive navigation to improve self-containment of multifunctional mobile robotic systems of radiological "MRK-2" and "RTK-08" surveillance (Research Institute of Robotics and Engineering Cybernetics, Saint-Petersburg).

The article [89] describes the problems of motion control of the mobile robot in the nondeterministic media by the potential field method. The review article [90] presents the analysis of research results in the field of mobile robot route planning based on the classical and evolutionary approach. The review of the control tasks without centralized coordination on the basis of the various algorithms is investigated in the article [91]. In the book [92] the modern methods of group control of mobile objects in uncertain environments are considered. The methods of group control using fuzzy logic and unstable modes are described.

The article [93] describes the main methods related to modular and reconfigurable mobile robotics. Structural modular morphology allows adaptation to the unstructured real environment. In comparison with crawler and wheeled robots, walking robots are more effective when moving along tough terrain. The article [94] considers the kinematic scheme of the executive mechanism of the octopod based on the biological prototype of the crab skeleton.

The aim of the paper [95] is to analyze the current state and development prospects of robotic systems for military purposes. The paper [96] investigates the problematic issues of the development of the autonomous ground-based robotic systems for special purposes for high-precision navigation.

Interaction of the mobile robots within the group is of great practical importance. The paper [97] is dedicated to modeling and analysis of collective biological systems in relation to the behavior of a swarm of mobile robots. The paper [98] presents an overview of the work on robotic research in development of swarm robotic systems for the real environment. The review of the papers on research of soft robotics with deformable bodies is given in the article [99]. In the papers [100] and [101] the review and prospects of research on the process of production of sensors and applications for soft biomimetical fish robots are considered.

2.9.3 Review of the Analogues

We also note some analogue use of the developed mobile robotic system. Four-wheeled mini robot Nerva LG (weight 4 kg) of company Nexter Robotics (France) is designed for remote surveillance of the environment [102]. Multipurpose compact FirstLook robots (weight 2.4 kg), SUGV (13.5 kg), PackBot (weight 18 kg) are developed by iRobot company (USA) for the tasks of countering self-made explosive devices and detection of potential hazards during military operation execution [103].

Throwable multipurpose mini robots are sufficiently relevant. Mini robot Yule (weight 0.6 kg) is developed by the Central Research Institute of Robotics and Engineering Cybernetics (Russia) [104]. It can conduct discreet remote audio-video observance in the field and urban infrastructure conditions. Throwable three-wheeled mini robot Pocketboot (0.85 kg) of Novatig (Switzerland) is designed for remote surveillance of the environment [103].

Small-scale four-wheeled robot MTGR (weight 9.4 kg) is developed by Roboteam (Israel) and designed for mine clearance of explosive objects, surveillance, detection of chemical, bacteriological or radiation contamination. Compact robot on the tracked chassis DAGU (weight 10 kg) is developed by General Robotics (Israel) and designed for surveillance, assault operations. Walking robot CLOOS (weight 69.4 kg), is manufactured by the company Autonomous Solutions Inc. (USA) to search for mines and inspection of hazardous materials [103].

The mobile robotic systems "Cross-country vehicle TMZ" (weight 35 kg) and "Varan" (weight 185 kg) are designed for remote visual surveillance, search and disposal of explosive devices; they are developed by the Research Institute of Special Engineering of Moscow State Technical University after N.E. Bauman [105]. Robotic platform Warrior (weight 226.8 kg), iRobot (USA) is designed to move potentially hazardous objects, clearing the path, extinguishing fire or surveillance [104]. Multifunctional robotic crawler system "Platform-M" (weight 800 kg), is developed by the Scientific Research Institute of Technology; "Progress" is designed for surveillance and combat operations [103].

2.9.4 Robot Structure

The structure of the developed robotic system in the assembly is shown in Fig. 2.9.1. The base platform of the system has the following characteristics: length - 360 mm, width - 225 mm, height - 200 mm, continuous work time without recharging - 1 hour, maximum motion speed - 10 km/hour, minimum speed - 1cm/sec, weight - 3 kg.

The block diagram of the mobile robotic system with a three-level control system is shown in Fig. 2.9.2 [106, 107]. The lower level consists of four wheels, electric motors with encoders, motor power drivers and servo control microcontrollers. Microcontrollers function as adaptive proportional-integral-differential controllers Arduino Nano.

The mid-level microcontroller collects and pre-processes environmental information. The following units are connected to the mid-level

Fig. 2.9.1 Structure of the mobile robotic system.

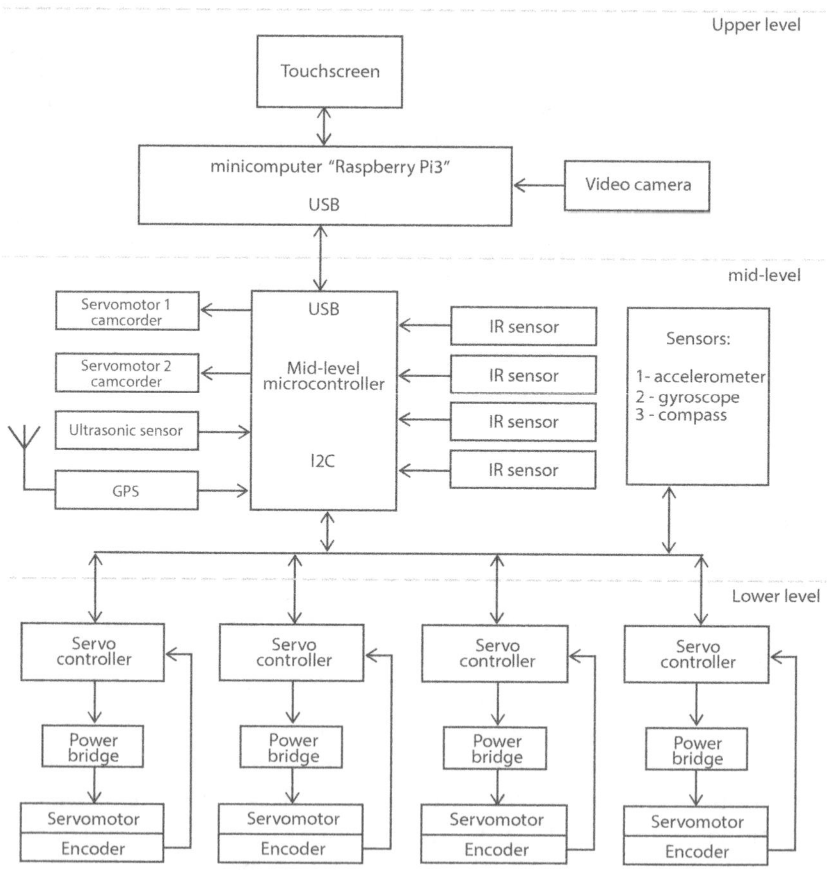

Fig. 2.9.2 Block diagram of the mobile robotic system.

microcontroller: accelerometer; gyroscope; compass; four infrared distance sensors; ultrasonic distance sensor and GPS sensor. At this level, there are also directing servo drives and cameras that are guided by this microcontroller.

The upper level controls the robot and is based on Raspberry PI3 minicomputer [108]. The minicomputer implements data transfer protocols: Bluetooth, Ethernet, Wi-Fi. The robot camera gives the possibility of implementing artificial vision; all the levels are interconnected by data communications protocols. The mid-level microcontroller is associated with the upper-level microcomputer by UART protocol (Universal asynchronous receiver/transmitter). This connection allows avoiding malfunctions at different signal levels. The mid-level and lower-level microcontrollers are interconnected by protocol I2C or TWI, this provides minimum data exchange rate.

2.9.5 Modeling Concept

Let us proceed to development of the computer three-dimensional model of the robotic system and its outer body lines based on the operating conditions. It is necessary to develop variants of the robot body with the upper platform, where the cargo compartment or the arm of manipulator will be located. The authors have experience in solving industrial design tasks in the field of mechanical engineering and hydroaviation [70, 109].

First, the schematic structural design of the robot's internal platform is carried out on the basis of the overall dimensions and layout design of the components. Further, in the process of searching for the lines of the outer body of the robot, the associative series of biological prototypes was analyzed, and some visual solutions were found [39, 41]. The Figure 2.9.3 shows variants of the visual search based on the arthropod crayfish and crabs. Later on the set of sketches will determine the style and color scheme of the future product; red mangrove crab Sesarma [110, 111] was chosen as a biological prototype. The crab is a representative of arthropods, a class of crustaceans; it has 5 pairs of limbs, the front pair has the form of powerful claws. With the loss of one of them, a new claw overgrows, so often in nature claws are asymmetric. Because of the overall claws crabs usually move backwards or sideways.

2.9.6 Modeling Stages

As a result of the stage-by-stage search taking functionality and aesthetics into account, one of the variants of the volumetric solution closely meeting

Fig. 2.9.3 Variants of visual searching for the body lines.

the requirements of the technical specification was worked out (Fig. 2.9.4, below).

The body shall be not only aesthetic, but also practical, preventing ingress of the foreign objects and dust into the inner part of the robot. In addition to the above tasks, the body plays the role of the bearer of

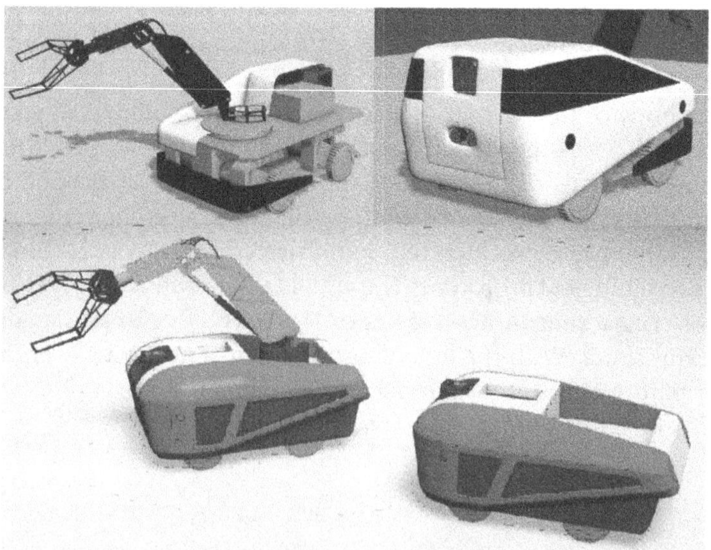

Fig. 2.9.4 Three-dimensional search models of the future body.

identity – corporate identity. The body contains visual information about the developer's company; it can be painted in corporate colors, or acquire camouflage color for some special tasks.

The final version of the body lines is dictated by the modern trends in the transport design and ergonomics. Plastic material is used for the body; gloss coating of the material can be replaced by more practical – matte one. The body structure consists of four parts (Fig. 2.9.5). The central part forms the frontal part of the body and the platform for the manipulator. Two symmetrical halves on the right and left, have the holes for infrared sensors; the rear lower cover closes down the wheels partly.

At the next development stage, a three-dimensional computer model of the body prototype is created. The variety of software products of two-dimensional and three-dimensional graphics allows implementing various design ideas. First, you need to have two-dimensional drawings, diagrams or photos, which are then exported to the universal formats for further processing in three-dimensional space. DWG is one of these formats; we will use the graphics system 3ds Max [54, 71] as a program of three-dimensional modeling. This software product along with other methods allows creating models through High-Poly (high polygonal)

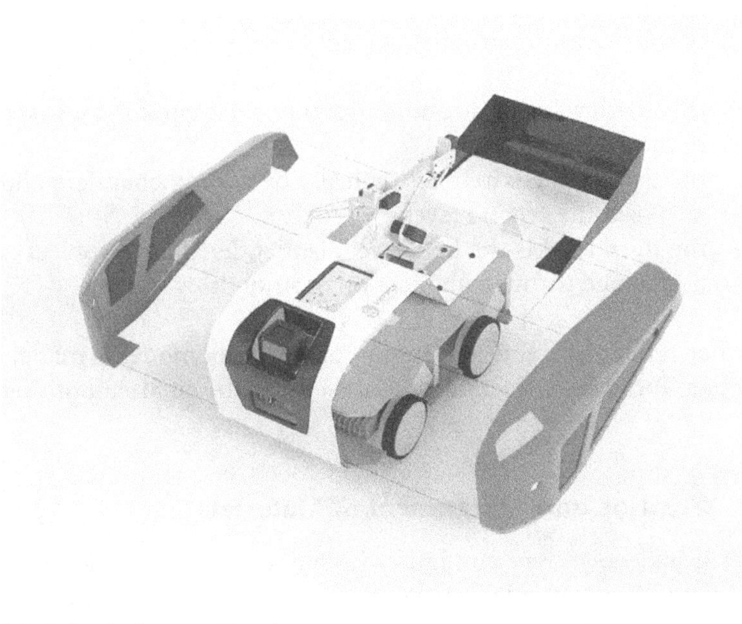

Fig. 2.9.5 Robot body assembly scheme.

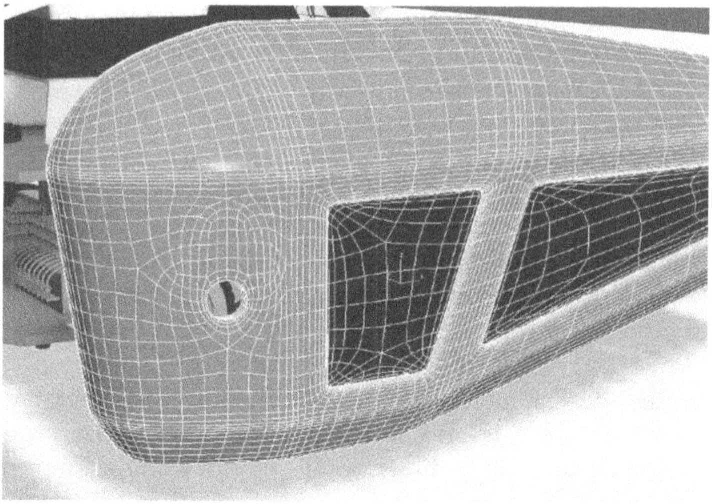

Fig. 2.9.6 High polygonal grid of body model.

modeling. Due to this the lines of the model will not have the faceted view, but a more streamlined shape.

To create a realistic model of the body in the graphic system 3ds Max the projections, cuts and sections of the future body are used. The standard scheme of High-Poly modeling consists of gradual increasing in the level of detail of the three-dimensional object:

- the first level is basic and represents the approximate shape of the object;
- the base shape is refined typically by adding chamfers and curves at the second level;
- the third level is the final one; distinct detail of the object is performed through application of smoothing plugins.

Further refinement and fine tuning of the body model is made; at the final stage TurboSmooth modifier is used for the final smoothing (Fig. 2.9.6).

2.9.7 Creation and Assignment of Materials

Further it is necessary to configure the model to create and apply materials. The methods of creating the materials are determined by the subsequent method of scene rendering. The external rendering unit V-Ray from ChaosGroup will be applied to create realistic rendering in the 3ds Max graphics system. This software package contains many parameters allowing controlling the quality of scene rendering and countdown time. The renderer requires the use and configuration of its own set of VRayMtl materials, which is different from standard materials.

The material setting is carried out in the basic parameters roll of the material editor. When creating a scene with a robot model, several materials with characteristic parameters were applied. Before assigning the materials, divide the three-dimensional model into the component elements. Then create the material to simulate plastic and apply it to the appropriate elements.

Let's consider the main parameters of the created material (Fig. 2.9.7):

- Diffuse roll is the color of the surface dispersion, diffuse color or just the basic color of the object surface. To do this you can choose a simple solid color or use texture. In our case, we choose without texture, imitating white plastic.
- Next roll is Reflection adjusting the reflective properties of the material. Surfaces in the surrounding space have different reflective properties. If 100% black color is selected in this window, it deprives the surface of the material of reflections; white, on the contrary, makes it completely reflective. All the intermediate values of grey color determine reflection intensity.

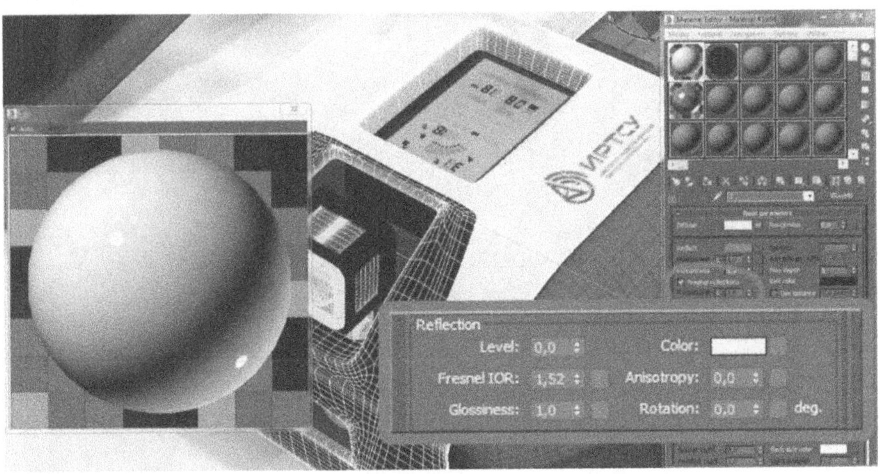

Fig. 2.9.7 Dialog box for setting of the material reflection.

- It is necessary to pay attention to the following parameter of setting of reflection and transmission of Fresnel IOR (index of refraction); this is the Fresnel's coefficient which depends on refraction of light rays.

The rest of the materials in this scene are set up in the same way. Separately it is worth noting the setting of the material for the inscriptions on the robot body. In this case, the texture based on the bitmap image of the company logo is applied as diffuse color in the Diffuse roll. Assignment of materials occurs at the level of polygons, with preliminary allocation of the corresponding group.

2.9.8 Lighting Installation and Rendering

At the next stage the adjustment, setting of lighting parameters with subsequent rendering is performed. Graphic modeling systems use two types of lighting: artificial (lamps) and natural (sky and sun light). The most realistic renderings are obtained by combining them; this lighting option is called mixed or combined.

The used renderer VRay is quite popular software, different from the analogues with quality and speed. For our stage we will simulate studio lighting (artificial). To do this place the robot model on the base without a clear horizon line (the base plane smoothly passes into the background plane). Then we get a soft background on which the created robot model is located (Fig. 2.9.8).

Modeling of the artificial lighting is carried out by installing of two equivalent VRay-Light sources from the opposite sides. This allows getting

Fig. 2.9.8 Creation of environment scene for rendering.

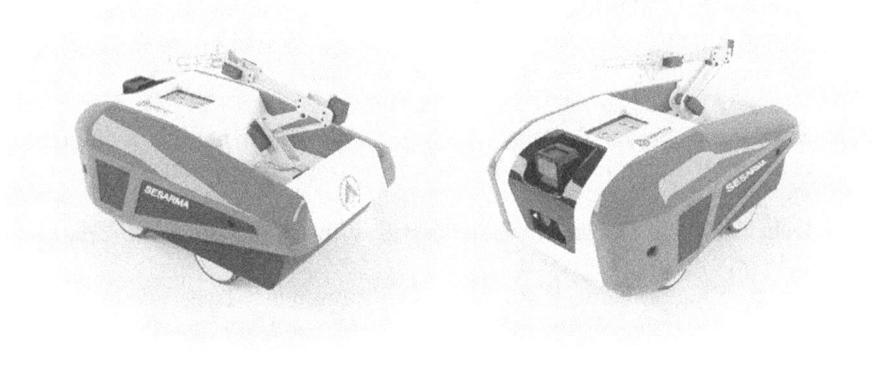

Fig. 2.9.9 Final scene rendering.

glare on the model surface necessary for realistic rendering. To adjust the shooting angle, a camera is installed; the results of the final scene rendering with the robot model are shown in Fig. 2.9.9.

2.9.9 Conclusion

In conclusion, it may be noted that in this paper the design was described and computer modeling of the developed mobile robotic system was carried out [112]. It can be used both for industrial and educational purposes. As a result of modeling, aesthetic and functional body was created, which emphasizes the main purpose of this product. The process of the preliminary prototyping makes possible rational selection and justification of the design of the robotic system before its production.

References

1. Petrov, G.F., *Amphibious and ekranoplanes of Russia in 1910-1999*, 248p, M. Rusavia, 2000.
2. The official website of the Beriev Aircraft Company. URL: www.beriev.com, cited 16 May 2014.
3. Panatov, G.S. and Udalov K.G., *The Illustrated Encyclopedia of Aircraft Beriev, 1932-1945*, 280p, Aviko Press, Moscow, 1988.
4. Panatov, G.S., *Beriev Aircraft, 1945-1968*, 223p, Restart +, Moscow, 2001.
5. Panatov, G.S., *Beriev Aircraft, 1968-2002*, 256p, RA Intervestnik, Moscow, 2002.
6. Hammond, D., Graphics in conceptual aircraft design - A designer's viewpoint, 1986. http://arc.aiaa.org/doi/abs/10.2514/6.1986-2733.

7. Haimes, R. and Drela M., On the construction of aircraft conceptual geometry for high-fidelity analysis and design. *50th AIAA Aerospace Sciences Meeting Including the New Horizons Forum and Aerospace Exposition*, Article numberAIAA 2012-0683, USA, 9–12 January 2012, 2012.
8. Xiao, G.-M., Feng, Y., Tang, W., Gui, Y.-W., Aerodynamics configuration conceptual design for ATLLAS-M6 analog transport aircraft. *Acta Aerodyn. Sin.*, 30, 5, 592–596, 2012.
9. Liebeck, R.H., Design of the Blended Wing Body Subsonic Transport. *J. Aircr.*, 41, 1, 97–104, 2004.
10. Saeed, T.I., Graham, W.R., Hall, C.A., Boundary-layer suction system design for laminar-flying-wing aircraft. *J. Aircr.*, 48, 4, 1368–1379, 2011, doi: 10.2514/1.C031283.
11. McMasters, J.H. and Cummings R.M., Airplane Design - Past, Present and Future. *J. Aircr.*, 39, 1, 10–17, 2002.
12. Sanders, B., Joo, J.J., Reich, G.W., Conceptual skin designs for morphing aircraft. *16th International Conference on Adaptive Structures and Technologies*, pp. 275–281, 2006.
13. Darwish, F.H., Atmeh, G.M., Hasan, Z.F., Design analysis and modeling of a general aviation aircraft. *Jordan J. Mech. Ind. Eng.*, 6, 2, 183–191, 2012.
14. Sarakinos, S.S., Valakos, I.M., Nikolos, I.K., A software tool for generic parameterized aircraft design. *Adv. Eng. Softw.*, 38, 39–49, 2007.
15. Azamatov, A., Lee, J.-W., Byun, Y.-H., Comprehensive aircraft configuration design tool for Integrated Product and Process Development. *Adv. Eng. Softw.*, 42, 35–49, 2011.
16. Abbasov, I.B., *Basics of three-dimensional modeling in the graphics system 3 ds Max 2009*, 176p, DMK Press, Textbook. Moscow, 2009.
17. Abbasov, I.B., Computational modeling of amphibious Be-200. *Proceedings of the Southern Federal University. Technical sciences*, pp. 160–164, 2009.
18. Abbasov, I.B. and Orekhov V.V., *Amphibious. Computational modeling*, 69p, LAP Lambert Academic Publishing, Saarbrucken, Germany, 2012, (www.lap-publishing.com).
19. Raymer, D.P., *Living in the Future; The Education and Adventures of an Advanced Aircraft Designer*, 360p, Design Dimension Press, Los Angeles, 2009.
20. Yeger, S.M., Matvienko, A.M., Shatalov, I.A., *Basics of aircraft: Textbook*, 720p, Mashinostroenie, M, 2003.
21. Abbasov, I.B. and Orekhov V.V., Computational modeling of multipurpose amphibious aircraft Be-200. *Adv. Eng. Softw.*, 69, 3, 12–17, 2014, doi: http://dx.doi.org/10.1016/j.advengsoft.2013.12.008.
22. Abbasov, I.B. and Orekhov V.V., Computational modeling of passenger amphibian aircraft Be-200 cabin interior. *Adv. Eng. Softw.*, 76, 154–160, 2014, doi: http://dx.doi.org/10.1016/j.advengsoft.2014.07.003.
23. Brooker, P., Civil aircraft design priorities: air quality? climate change? noise? *Aeronaut. J.*, 110, 517–532, 2006.

24. Vink, P. and Brauer K., *Aircraft Interior Comfort and Design (Ergonomics Design Management: Theory and Applications)*, First ed., 128p, CRC Press, Boca Raton, 2011, http://refhub.elsevier.com/S2212-540X(13)00016-3/sbref3.
25. Hall, A., Mayer, T., Wuggetzer, I., Childs, P.R.N., Future aircraft cabins and design thinking: Optimisation vs. win-win scenarios. *Propul. Power Res.*, 2, 2, 85–95, 2013, doi: http://dx.doi.org/10.1016/j.jppr.2013.04.001.
26. Gudmundsson, S., *General Aviation Aircraft Design*, pp. 521–545, Elsevier, 2014, Chapter 12. The Anatomy of the Fuselage.
27. Howe, D., *Aircraft Conceptual Design Synthesis*, 474p, Professional Engineering Pub. Ltd, London, 2000, http://dx.doi.org/10.1016/j.advengsoft.2013.12.008.
28. Orekhov, V.V. and Abbasov I.B., Computational modeling of the interior of salon the passenger modification of amphibious Be-200. *Proceedings of the Southern Federal University. Technical sciences*, pp. 110–116, 2012.
29. W.A.J. and Anemaat, Conceptual Airplane Design Systems. Vehicle Design, Air Vehicle Design, Published Online: 2010. http://dx.doi.org:10.1002/9780470686652.eae394.
30. The official website. URL: http://www.airwar.ru, cited 16 May 2014.
31. Orekhov, V.V. and Abbasov I.B., Computational modeling of amphibious Be-103. *Proceedings of the Southern Federal University. Technical sciences*, 1, pp. 121–125, 2011.
32. Bolsunovsky, A.L., Sonin, O.V. *et al.*, Flying wing - problems and decision. *Aircraft Design*, 4, 4, 193–219, 2001, doi: http://dx.doi.org/10.1016/S1369-8869(01)00005-2.
33. Gavel, H., Berry, P., Axelsson, A., Conceptual design of a new generation JAS 39 gripen aircraft. *Collection of Technical Papers - 44th AIAA Aerospace Sciences Meeting*, Reno. USA, 9-12 January 2006, vol. 1, pp. 395–406, 2006.
34. Kroll, E., Condoor, S.S., Jansson, D.G., *Innovative Conceptual Design: Theory and application of parameter analysis*, 227p, Cambridge University Press, 2001.
35. Website/Internet resource. Mode of access www/URL: http://airwar.ru (date access 20.07.2016).
36. Abbasov, I.B., Conceptual model of aircraft "Chiroptera. *Am. J. Mech. Eng.*, 2, 2, 47–49, 2014, doi: 10.12691/ajme-2-2-3.
37. Jenkinson, L.R. and Marchman J.F., *Aircraft design projects*, 371p, Butterworth-Heinemann, Oxford, 2003.
38. Runge, V.F. and Manusevich Y.P., *Ergonomics in environmental design*, 328p, Architecture-C, Moscow, 2005.
39. Happian–Smith, J., *An Introduction to Modern Vehicle Design*, 600p, Elsevier Limited, 2002.
40. Website/Internet resource. Mode of access www/URL: https://en.wikipedia.org/wiki/Northern_lapwing (date access 19.05.2016).
41. Vasin, S.A., Talaschuk, A.U. *et al.*, *Design and modeling of industrial products*, 692p, Mashinostroenie, Moscow, 2004.
42. Abbasov, I.B., *Computational modeling in industrial design*, 92p, DMK Press, Moscow, 2013.

43. Winzen, J. and Marggraf-Micheel C., Climate preferences and expectations and their influence on comfort evaluations in an aircraft cabin. *Build. Environ.*, 64, 146–151, 2013. https://doi.org/10.1016/j.buildenv.2013.03.002.
44. Filippone, A., Aircraft noise prediction. *Prog. Aerosp. Sci.*, 68, 27–63, 2014, doi: https://doi.org/10.1016/j.paerosci.2014.02.001.
45. Haddad, R., Research and Methodology for Interior Designers. *Procedia - Soc. Behav. Sci.*, 122, 283–291, 2014, doi: https://doi.org/10.1016/j.sbspro.2014.01.1343.
46. Cabin Design & Interiors/Website/Internet resource, Mode of access. www/URL: http://www.futuretravelexperience.com/up-in-the-air/cabin-design-interiors/(date access 29.04.2017).
47. Sforza, P.M., *Commercial Airplane Design Principles*, 1st Edition, pp. 47–79, Chapter 3, Fuselage Design, Butterworth-Heinemann, Oxford, 2014, https://doi.org/10.1016/B978-0-12-419953-8.00003-6.
48. Laananen, D.H., Crashworthiness analysis of commuter aircraft seats and restraint systems. *J. Safety Res.*, 22, 2, 83–95, 1991. https://doi.org/10.1016/0022-4375(91)90016-O.
49. Brindisi, A. and Concilio A., Passengers Comfort Modeling Inside Aircraft. *J. Aircr.*, 45, 6, 2001–2008, 2008, doi: https://doi.org/10.2514/1.36305.
50. Ahmadpour, N., Lindgaard, G., Robert, J.-M., Pownall, B., The thematic structure of passenger comfort experience and its relationship to the context features in the aircraft cabin. *Ergonomics*, 57, 6, 801–815, 2014, doi: http://dx.doi.org/10.1080/00140139.2014.899632.
51. Hiemstra-van Mastrigt, S., Groenesteijn, L., Vink, P., Kuijt-Evers Lottie, F.M., Predicting passenger seat comfort and discomfort on the basis of human, context and seat characteristics: a literature review. *Ergonomics*, 60, 7, 889–911, 2017. http://dx.doi.org/10.1080/00140139.2016.1233356.
52. Abbasov, I.B. and Orekhov, V.V., Conceptual Model of "Lapwing" Amphibious Aircraft. *Mech. Mater. Sci. Eng. J.*, 7, 209–221, 2016, doi: 10.13140/RG.2.2.12856.14081. http://mmse.xyz/en/machine-building-vol-7/.
53. Eckert, C. and Stacey M., Sources of inspiration: A language of design. *Design Stud.*, 21, 5, 523–538, 2000.
54. Mooney, T., *3ds Max speed modeling for 3D artists*, 422p, Packt Publishing, 2012.
55. Yun, L., Bliault, A., Doo, J., *WIG craft and Ekranoplan: Ground Effect Craft Technology*, 450 p, Springer, New York, 2010.
56. Ekranoplan, VOLGA2/Website/Internet resource. Mode of access (date access 21.07.2018) www/URL: http://www.ikarus342000.com/VOLGA2page2.htm.
57. Ekranoplan, AQUAGLIDE-5"/Website/Internet resource. Mode of access (date access 19.07.2018) www/URL: http://www.attk-invest.com/Eng/product/Aq-5.htm.
58. Ekranoplan, CYG-11 aircraft/Website/Internet resource. Mode of access (date access 21.07.2018) www/URL: https://defence.ru/article/1407/.

59. Ekranoplan, AirFish 8/Website/Internet resource. Mode of access (date access 23.07.2018) www/URL: http://www.wigetworks.com/airfish-8/.
60. Ekranoplan, HOVERWING 20/Website/Internet resource. Mode of access (date access 21.07.2018) www/URL: http://www.flightboat.net/.
61. Ekranoplan, Flarecraft/Website/Internet resource. Mode of access (date access 21.07.2018) www/URL: https://kimrobertsmarine.com/.
62. Ekranoplan, ARON-7, Korea WIG craft/Website/Internet resource. Mode of access (date access 21.07.2018) www/URL: https://defence.ru.
63. Nebylov, A.V. and Wilson, Ph., *Ekranoplane - Controlled Flight Close to Sea*, 300p, WIT-Press/Computational Mechanics Publications, Monograph. Southampton, UK, 2001.
64. Yang, Z. and Yang W., Complex Flow for Wing-in-ground Effect Craft with Power Augmented Ram Engine in Cruise. *Chinese J. Aeronaut.*, 23, 1, 1–8, 2010. https://doi.org/10.1016/S1000-9361(09)60180-1.
65. Mobassher, T., Maimun, A., Ahmed, Y., Jamei, S., Priyanto, A., Rahimuddin, Experimental Investigation of a Wing-in-Ground Effect Craft. *Scientific WorldJournal*, l, 7 pages, 2014. Article ID 489308https://doi.org/10.1155/2014/489308.
66. Rozhdestvensky, K.V., Wing-in-ground effect vehicles. *Prog. Aerosp. Sci.*, 42, 3, 211–283, doi: 10.1016/j.paerosci.2006.10.001.
67. Hameed, H., The design of a four-seat reverse delta WIG craft. *Maldives Natl. J. Res.*, 6, 1, 7–28, 2018, doi: 10.13140/RG.2.2.11122.61129. February.
68. Wang, H., Teo, C.J., Khoo, B.C., Goh, C.J., Computational Aerodynamics and Flight Stability of Wing-In-Ground (WIG) Craft. *Procedia Eng.*, 67, 15–24, 2013.
69. Andersen, N.M. and Cheng L., The marine insect Halobates (Heteroptera: Gerridae): Biology, Adaptations, Distribution and Phylogeny. *Oceanogr. Mar. Biol.: An Annual Review*, 42, 119–180, 2004, doi: 10.1201/9780203507810.ch5.
70. Abbasov, I.B. and Orekhov V.V., Computational modeling of the cabin interior of the conceptual model of amphibian aircraft "Lapwing. *Adv. Eng. Softw.*, 114, 227–234, 2017, doi: http://dx.doi.org/10.1016/j.advengsoft.2017.07.003.
71. Abbasov, I.B., *A Fascinating Journey into the World of 3D Graphics with 3ds Max*, 239 p, Amazon Digital Services LLC, 2017, ASIN: B076333ZQD [Electronic resource] https://www.amazon.com/Fascinating-Journey-into-World-Graphics-ebook/dp/B076333ZQD/ref=asap_bc?ie=UTF8.
72. Abbasov, I.B. and Orekhov V.V., Conceptual Model and Interior Design "Water Strider" Ekranoplan. *Int. Rev. Mech. Eng.*, 13, 3, 162–172, 2019, doi: https://doi.org/10.15866/ireme.v13i3.16244.
73. OAO Central Design Bureau for hydrofoil vessels named after R.E. Alekseev. "http://www.ckbspk.ru/.
74. Lukomsky, Yu.A. and Starichenkov, A.L., Prediction of the stability of the movement of vessels with dynamic principles of maintenance. News *SPbGETU LETI*, 6, 13–17, 2004.

75. Shamim, M., The applicability of hydrofoils as a ship control device. *J. Mar. Sci. Appl.*, 14, 1–6, 2015, doi: 10.1007/s11804-015-1314-x.
76. Lyakhovitsky, A.G., Sakhnovsky, E.B., Sakhnovsky, B.M., Design of high-speed catamarans with hydrofoils. *Shipbuilding*, 2, 759, 9–15, 2005.
77. Aktas, B., Turkmen, S., Sasaki, N., Atlar, M., Turnbull, M., Knos, M., A study on hydrofoil application to assist wind farm support activity of a catamaran. *Conference: 10th Symposium on High-Performance Marine Vehicles*, 20p, 2016.
78. Bolotin, A.A., Application of the discrete vortex method for the study of underwater wings. *Proceedings of NSTU*, vol. 3, pp. 209–213, 2015.
79. Gribov, K.V. and Fedoreev G.A., High-speed vessels on the wings for the Far Eastern Basin. *B. Eng. School of the Far Eastern Federal University*, 1, 30, 98–115, 2017.
80. Korzinov, N., Inspired: from "Swallows" to "Dolphin. *Pop. Mech.*, 12, 2007. https://www.popmech.ru/technologies/7176-okrylennye-ot-lastochki-do-delfina/#part0.
81. Besnard, E., Schmitz, A., Kaups, K., Tzong, G., Hefazi, H., Kural, O., Chen, H., Cebeci, T., *Hydrofoil design and optimization for fast ships*, vol. 56, 61p, American Society of Mechanical Engineers, Aerospace Division (Publication) AD, 1998.
82. Sokolov, V.P., New types of high-speed passenger ships for the rivers of Siberia and the Far East. *Transport of the Russian Federation*, 1, 56, 64–66, 2015.
83. Maciej, R. and Bednarek A., The experimental studies on hydrofoil resistance. *Arch. Civ. Mech. Eng.*, 7, 3, 167–175, 2007, doi: 10.1016/S1644-9665(12)60024-7.
84. Karlsson, S. and Urde J., *Hydrofoiling Europe-Dinghy Development of separate hydrofoils as complement to Europe-dinghies*, 120p, Lund University, 2018.
85. Eickmeier, J., Dalanaj, M., Gray, J., Kotecki, M., *OCE Hydrofoil Development Team Spring/Summer*, 85p, Florida Institute of Technology, 2006.
86. Electric hydrofoil finally ready to skim the waves MARINE Stu Robarts December 15th, 2016. https://newatlas.com/quadrofoil-q2-electric-hydrofoil/46973/.
87. Vasiliev, A.V., Mobile mini-intelligence robots: Current status, characteristics and general development trends. *Izv. SFedU. Technical science*, 3, 119–124, 2010.
88. Lopota, A.V., Polovko, S.A., Smirnova, E.Yu., Plavinsky, M.N., Main results and promising areas of research in the field of navigation and control of mobile robotic complexes. *Res. Sciencecity*, 2, 4, 49–53, 2013.
89. Soloviev, V.V., Pshikhopov, V.K., Shapovalov, I.O., Finaev, V.I., Beloglasov, D.A., Planning of the mobile robot motion in nondeterministic environments with potential fields method. *Int. J. Appl. Eng. Res.*, 10, 21, 41954–41961, 2015.

90. Raja, P. and Pugazhenthi S., Optimal path planning of mobile robots: A review. *Int. J. Phys. Sci.*, 7, 9, 1314–1320, 2012. http://dx.doi.org/10.5897/IJPS11.1745.
91. Bayındır, L., A review of swarm robotics tasks. *Neurocomputing*, 172, 292–321, 2016, doi: http://dx.doi.org/10.1016/j.neucom.2015.05.116.
92. Beloglazov, D.A. *et al.*, *Group management of mobile objects in uncertain environments*, V.H. Pshihopov (Ed.), 305p, Fizmatlit, M, 2015.
93. Moubarak, P. and Ben-Tzvi P., Modular and reconfigurable mobile robotics. *Rob. Auton. Syst.*, 60, 12, 648–1663, 2012. http://dx.doi.org/10.1016/j.robot.2012.09.002.
94. Kovalchuk, A.K., The choice of kinematic structure and the study of the dynamics of the tree-like actuator of the robot crab. *Mashinostroenie*, 7, 73–79, 2013.
95. Makarenko, S.I., Military robotic systems - current state and development prospects. *Control Systs. Commun. Secur.*, 2, 73–132, 2016.
96. Lapshov, V.S., Noskov, V.P., Rubtsov, I.V., Rudianov, N.A., Gurdzhi, A.I., Ryabov, A.V., Khrushchev, V.S., Prospects for the development of autonomous ground-based robotic complexes of a special military purpose. Izvestiya SFU. *Technical Science*, 1, 156–168, 2016.
97. Brambilla, M., Ferrante, E., Birattari, M., Dorigo, M., Swarm Intell Swarm robotics: A review from the swarm engineering perspective. *Swarm Intelligence*, 7, 1, 1–41, 2013, doi: http://dx.doi.org/10.1007/s11721-012-0075-2.
98. Barca, J.C. and Sekercioglu Y.A., Swarm robotics reviewed. *Robotica*, 31, 345–359, 2013, doi: http://dx.doi.org/10.1017/S026357471200032X.
99. Rus, D. and Tolley M., Design, Fabrication and Control of Soft Robots. *Nature*, 7553, 467–475, 2015. http://dx.doi.org/10.1038/nature14543.
100. Liu, H., Tang, Y., Zhu, Q., Xie, G., Present research situations and future prospects on biomimetic robot fish. *Int. J. Smart Sensing Intell. Syst.*, 7, 2, 458–480, Jun. 2014. http://dx.doi.org/10.21307/ijssis-2017-665.
101. Liu, G., Wang, A., Wang, X., Liu, P., A Review of Artificial Lateral Line in Sensor Fabrication and Bionic Applications for Robot Fish. *Appl. Bionics. Biomech.*, 2016, 15 pages, 2016, doi: https://doi.org/10.1155/2016/4732703. Article ID 4732703.
102. Nexter's Nerva LG Robot, The site of the company. [Official site]. URL: http://www.nexter-group.fr/nexter/Flipping_Book/Export_GB/#61/z/(access date:07.11.2018).
103. Military robots and their developers. [Official site]. URL: https://habr.com/company/smileexpo/blog/408731/(access date: 06.11.2018).
104. Military Review, Internet journal. [Official site]. URL: http://topwar.ru/66967-nazemnye-robotyot-zabrasyvaemyh-sistem-do-bezlyudnyh-transportnyh-kolonn-chast-1.html/(access date: 12.11.2018).
105. Department SM4-1, Design of structures of robotic systems [Ofits. site]. URL: http://www.niism.bmstu.ru/otdelyi-nii-sm/sm4-1/(access date: 14.11.2018).

106. Yurevich, E.I., *Basics of robotics*, Textbook for universities, 2nd ed., 401p, BHV-Petersburg, Pererab. and add. SPb, 2005.
107. Siegwart, R., Nourbakhsh, I.R., Scaramuzza, D., *Introduction to Autonomous Mobile Robots*, 488p, MIT Press, 2011.
108. Leshchenko, V.V., Leshchenko, S.A., Ignatiev, V.V., Spiridonov, O.B., The control program of the transport robot to automate the distribution of goods in stock: Application number 2016612284 from 03.17.2016. Certificate of state registration of computer programs №2016616217 from 06.06.2016.
109. Abbasov, I.B. and Gabrilyan H., Conceptual Design of "Lotos" Motorcar. *IOSR J. Comput. Eng.*, 18, 2, Ver.4, 33–36, 2016, doi: 10.9790/0661-1802043336. http://www.iosrjournals.org/iosr-jce/pages/18(2)Version-4.html.
110. Zarenkov, N.A., *Zoology of invertebrates. Arthropods. Crustaceans*, 304p, Publishing Lenand, M, 2015.
111. Koenemann, S. and Jenner R.A., *Crustacea and Arthropod Relationships*, CRC Press, 2005. 423p.
112. Abbasov, I.B., Ignatyev, V.V., Orekhov, V.V., Autonomous mobile robotic system "Sesarma. *IOP Conf. Series: Materials Science and Engineering*, vol. 560, 012001, 9p, 2019, doi: https://doi:10.1088/1757-899X/560/1/012001.
113. Menegon, L., da Silva, Vincenzi, S.L., de Andrade, D.F., Barbetta, P.A., Díaz Merino, E.A., Vink, P., Design and validation of an aircraft seat comfort scale using item response theory. *Applied Ergonomics*, 62, 216–226, 2017. https://doi.org/10.1016/j.apergo.2017.03.005
114. Ekranoplan, ESKA-1 (amphibian rescue craft)/Website/Internet resource. Mode of access (date access 18.07.2018) www/URL: http://www.airwar.ru/enc/xplane/eska.html.
115. Ekranoplan, "Burevestnik-24"/Website/Internet resource. Mode of access (date access 18.07.2018) www/URL: http://www.airwar.ru.
116. Yablonsky, P.P., Cruise ships of the fatherland, 98 p. WIG of the world, Moscow, 1997.
117. Ivanov, A.V., He was ahead of time: Rostislav Alekseev (acts of shipbuilders through the eyes of an aircraft engineer). 168p, N. Novgorod, Quartz, 2006.
118. Ji, B., Luo, X., Wu, Y., Peng, X., Duan, Y., Numerical analysis of unsteady cavitating turbulent flow and shedding horse-shoe vortex structure around a twisted hydrofoil. *Int. J. Multiphas. Flow*, 51, 33–43, 2013. doi.org/10.1016/j.ijmultiphaseflow.2012.11.008.
119. Caldwell, D., Caldwell, M., *The World of the Bottlenosed Dolphin*, Sokolov, A.S. (Ed.), 136p. M.: Gidrometeoizdat, 1980.
120. Abbasov, I.B., Orekhov, V.V., Computational modeling of the cabin interior of the conceptual model of amphibian aircraft "Lapwing". *Adv. Eng. Softw.*, 114, 227–234, 2017. http://dx.doi.org/10.1016/j.advengsoft.2017.07.003.

3

Development of Schemes of Multirotor Convertiplanes with Cryogenic and Hybrid Powerplants

Dmitriy S. Durov

Southern Federal University, Engineering Technological Academy, Department of Engineering Graphics and Computer Design, Taganrog, Russia

Abstract

The third chapter considers various types of aerodynamic configurations and possible areas of application of modern and perpective convertible aircrafts of short and vertical takeoff and landing, both land and ship-based. The possibilities of creating new types of rotary-wing high-speed air transport constructed on the basis of well-known and well-established designs of helicopters and airplanes are shown. The concepts and preli*minary tactical and technical requirements for heavy, medium and light convertible high-speed aircraft with cryogenic, liquefied natural gas driven, and hybrid diesel-electric and turboprop powerplants are presented.

Keywords: Convertiplane, plane, helicopter, rotorcraft, gyroplane, cryogenic hybrid powerplant, liquefied natural gas, electromotor-generator

3.1 Introduction

From the moment of the creation of the first aircraft heavier than air, and all along the way of their further improvement, the question of efficiency and the possibility of their use in various flight regimes has remained open. The growth of flight speeds and altitudes is associated with the deterioration of the takeoff and landing characteristics of the aircraft – an increase

---

Email: dsdurov@sfedu.ru

Iftikhar B. Abbasov (ed.) Computer Modeling in the Aerospace Industry, (137–160)

in the length of the takeoff run during takeoff and the inclease of runway while landing. The ability of aircraft to move quickly in space is important, but their takeoff and landing capabilities are no longer meeting modern requirements. The problem of using an aircraft from the water is partially solved by amphibious aircraft, vertical landing and takeoff or a short takeoff are available to helicopters. Currently, they are being replaced by convertible rotary-wing aircraft (CRWA) of vertical takeoff and landing, the mastering of which is a quite prospective task [1−4].

The efficiency of using helicopters is limited by the speed of their flight (300 km/h on average), so it becomes necessary to resort to higher-speed aircraft—convertible rotary-wing aircraft, known as converters and having the same aerodynamic configuration as regular aircraft. Convertiplanes can take off and land from any unprepared sites and a long runway is not needed.

3.2 Hydro Convertiplane is the New Opportunity for Modern Aviation

The high thrust-to-weight ratio of amphibious jet aircraft causes a certain range of problems. For their effective operation, it is required to exclude the ingress of exhaust gases of lifting engines into the air inlets in order to avoid reducing thrust and surging of the latter. During vertical takeoff and landing, bypass turbofan engines (BTE) due to the impact of hot jets of their gases create problems associated with erosion and destruction of the landing pad, which requires its durable thermal insulation coating [5]. The new approach to the development of ship platforms and any surface of the earth today is the use of not traditional jet aircraft, vertical takeoff and landing (VTOL) amphibians and amphibian helicopters, but more multifunctional hydro convertiplanes (HCP).

The modular approach to the HCP concept is attractive for creating on its basis a transport convertible rotary-wing aircraft used as a helicopter, an amphibious aircraft and a wing-in-ground-effect-craft. Combined with general approaches and reasons preventing the creation of a HCP, a number of convertiplane models constructed in the USA are being analyzed, where there is a considerable experience in their creation and operation [6].

One of the layouts of HCP (see Fig. 3.1) with the combination of helicopter and airplane flight modes is an aerodynamic high-profile scheme using three-beam tail and four tandem-mounted rotors in the rotary annular channels, as well as "composite wing" layouts, providing maximum protection for rotors and engines from ingress of sea water [7].

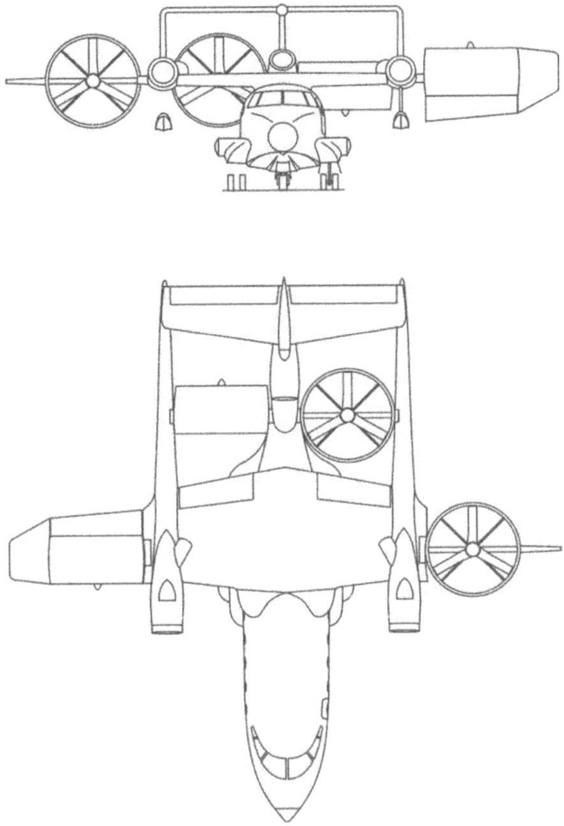

Fig. 3.1 Hydro convertiplane.

The arrangement shown in Fig. 3.2. may be used for land and naval applications.

The large thrust-to-weight ratio of the three-engine powerplant of HCP, combined with a decrease in the unit load on the power, significantly increases the service life and also makes it possible to distinguish its range of basic versions from the CRWA of naval aviation. For example, type 1 – search and rescue HCP, type 2 and type 3 – transport HCP (see Table 3.1).

The widespread use of convertiplanes for the needs of civil aviation is currently limited to contradictive requirements for their powerplant. The thrust on vertical takeoff, landing or hovering modes should be greater than the weight of the aircraft, and when flying at cruising speed, the thrust reaches only 18-22% of takeoff weight or 30-40% of available power.

Therefore, a large thrust-to-weight ratio of the convertible rotary-wing aircraft from 1.25 to 1.75 in its total weight inevitably increases the

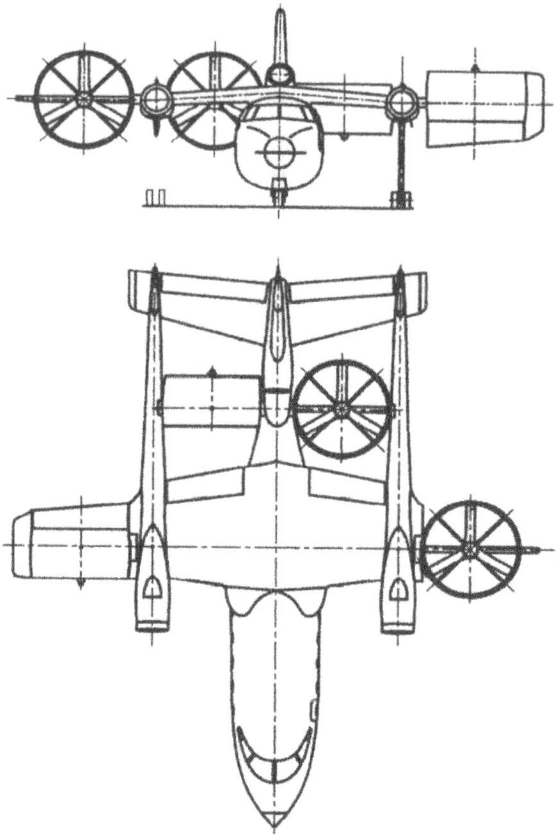

Fig. 3.2 Convertiplane.

proportion of the weight of the powerplant [8]. At the same time, the share of fuel and the time for which it is consumed decreases, which increases the specific fuel consumption.

Other factors hindering the use of convertible rotary-wing aircraft are the low relative thrust of their powerplants (PP), the low cruising flight speed of 450-600 km/h [8], their low profitability with considerable operating costs. To increase the efficiency of cruise flight at subsonic speeds, it is necessary to increase the relative thrust. The increased relative thrust allows making a cruise flight with more complete utilization of the PP thrust and with a lower specific fuel consumption. But with increasing cruise speed, aerodynamic drag increases and more horizontal thrust is required [9].

The creation of PP, having two MEs and one BTE, will significantly increase the speed of the HCP flight. Additional mechanically driven BTE

Table 3.1 Preliminary summary technical requirements for HCP.

No.	Parameters	Values		
		Type 1	**Type 2**	**Type 3**
1.	Dimensions:			
	Fuselage length, m	13.8	19.8	19.8
	Height at chassis, m	4.8	6.2	6.2
	wing span with annular channels, m	14.8	20.8	20.8
	wing area with annular channels, m^2	27.84	52.55	52.55
2.	Powerplant (PP):			
2.	main engine (ME), model	TV2-117A	TV7-117V	TV7-117V
2.1	total power, hp	1,500 × 2	2,800 × 2	2,800 × 3
2.2	additional engine (BTE), model	AL-55	AL-55	–
2.3	additional takeoff thrust, kgf	2,000 × 1	2,000 × 2	–
3.	Masses and loads:			
3.1	normal with vertical takeoff, kg	7,000	13,360	13,090
3.2	short takeoff, kg	8,600	16,360	16,090
3.3	normal payload, people/(t)	2 + 14 / (1.6)	2 + 28 / (3.0)	2 + 28 / (3.0)
3.4	empty, kg	3,950	7,810	7,540
4.	Fuel capacity, kg	1,450	2,550	2,550
5.	Diameter of three-bladed rotors, m	2.37	3.27	3.25
6.	Rotor disk area, m^2	17.63	33.57	33,16

(*Continued*)

Table 3.1 Preliminary summary technical requirements for HCP. (*Continued*)

No.	Parameters	Values		
		Type 1	Type 2	Type 3
7.	The rotation speed of three-bladed rotors: vertical lift, rpm cruise flight, rpm	 2,590 2,080	 2,080 1,670	 2,080 1,670
8.	Specific load on the rotor disk area, kgf/m^2	397.05	397.97	398.46
9.	Specific load on power, kgf/hp	1.56	1.56	1.56
10.	Specific load on the wing with a maximum takeoff weight according to p. 3.2, kgf/m^2	308.91	311.32	306.18
11.	Flight performance:			
	The 1st cruise speed at 40% of the available capacity of the ME, km/h	640	680	640
	The 2nd cruise speed on the airplane mode during the 1st ME + BTE operation, km/h	820	860	–
	maximum speed, km/h	840	880	690
	practical ceiling, m	6,000	8,000	8,000
	takeoff distance at running landing/ during takeoff with a short takeoff, m	120/80	150/100	180/120

transmitting most of the available power to the rotors in the hover modes, and provide an increase in the vertical lifting thrust. After disconnecting the BTE from rotors and only due to the joint thrust of the rotors and their reactive marching thrust on aircraft flight modes, there is a possibility to increase the flight speed of the HCP to 880 km/h (see Table 3.1, the comparison of Type 2 and Type 3).

A lift fan provides approximately 50% of the thrust required for hover flight [10]. Therefore, the creation for high-speed HCP (Type 1 and Type 2) of such a BTE with a deflectable nozzle, equipped with a mechanical drive, transferring part of the power from its rotor to the rotors in the rotary annular channels, will allow it to achieve efficiency of its single PP (Type 3). This is explained by the fact that an optimal ratio is established between the horizontal marching and vertical lifting rods by combining engines with good cost effectiveness (HCP + BTE) into a combined PP. At the same time, the HCP will perform cruising flight in the mode close to the maximum, and, therefore, it will have a rather low specific fuel consumption [9].

3.3 Peculiarities of Control of the Vertical Takeoff and Landing Aircraft in the Transitional and Hovering Mode

One of the directions of development of modern aviation has become vertical takeoff and landing aircrafts (VTOL aircrafts). Such aircrafts do not need extensive runways and can take off and land from limited sites and in cramped urban-based spaces [11].

The static rotor thrust can be increased by surrounding the rotors of the multiple-rotor VTOL aircrafts by annular channels [7, 11, 12]. Due to the annular channels, an intensive increase in wave resistance is postponed until the tip of the rotor blade ends reaches 280–290 m/s and the static thrust increases. Behind these rotors, an air jet is formed, which has great energy, which makes it possible to install efficient aerodynamic steering surfaces at the exit of the annular channels. The annular channels create additional bearing surfaces that increase the lift force of the wing and make it possible to increase safety, eliminating the possibility of contact of foreign objects with cantilever rotors. They also reduce noise exposure in comparison with rotors without annular channels. Channel rotors with steering surfaces at the exit of the annular channels simplify the longitudinal and transverse controls on the hover modes, as well as the course control. This will allow the use of rotors without cyclic step change and reduce the weight of the structure [6, 7, 13].

The aerodynamic control of VTOL aircrafts in the hover mode is not effective because of the low translational speed of the aircraft. VTOL aircraft requires additional aircrafts to create control accelerations and moments around axes connected with the aircraft, for example, to make a turn in the horizontal plane (Fig. 3.3). With the help of these control actions, the tasks of stabilization, maneuvering, balancing and flight control along the vertical are solved [11, 14].

When performing, for example, the right bank with a twin-rotor VTOL aircraft (see Fig. 3.4) the main rotor on the axis line receives a negative increment of the flow velocity $\Delta\bar{V}_r = -y_{rot}\dot{\Phi}$, respectively, the left main rotor receives a positive increment $\Delta\bar{V}_l = y_{rot}\dot{\Phi}$, here Φ – bank angle. In this case, the right main rotor receives a positive increment thrust ΔF_r, and the left main rotor receives negative increment $-\Delta F_l$. Thus, the rotational movement around the linear axis is counteracted by a moment of forces around the same axis, proportional to the angular velocity of rotation will be [11]:

$$\Delta L = -y_{rot}(\Delta F_r - \Delta F_l),$$

Where y_{rot} – distance from the longitudinal axis of a VTOL aircraft to the rotor axis.

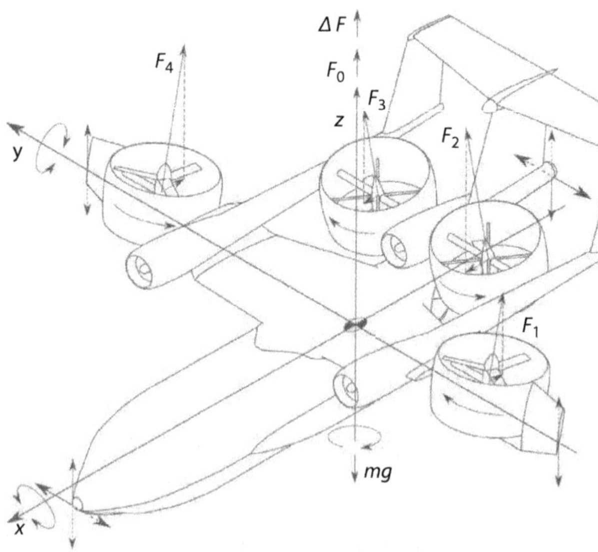

Fig. 3.3 Control forces and moments affecting VTOL aircrafts on the mode of vertical takeoff, landing and hovering.

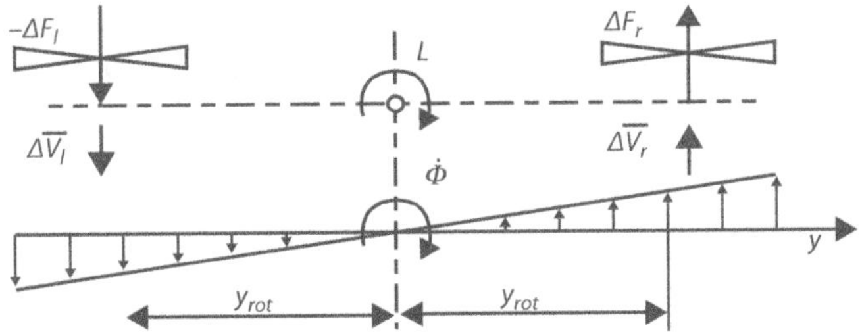

Fig. 3.4 Increment of the speed of the flow incident on the rotors during rotation of the VTOL aircraft around the longitudinal axis [11].

An important problem is the failure of the engine during the hovering mode of VTOL aircrafts, especially with eccentrically located engines. The question arises whether there will be enough vertical reserve for the remaining engines in order to maintain the hover mode. Total power N of the powerplant with n turboprop engines with power N_0 each is equal to:

$$N = n\,N_0.$$

The symmetry of the lifting forces due to the interconnection of the rotors through the shaft system can be maintained when one engine fails without the need to turn off the engine symmetrical to it, as in the case of VTOL with turbofan engines. The power rest N_{rest} divided by the normal total power N_Σ, taking into account the emergency power increase, characterized by the coefficient k in the event of a single engine failure [11], is reported to be equal to:

$$\frac{N_{rest}}{N_\Sigma} = \frac{n-1}{n}k.$$

When a single engine fails, the thrust generated by the rotors decreases to a lesser extent than the remaining power of the remaining engines, as the rotor of the failed engine, due to the transmission connecting the rotors of all the engines, rotates and creates lifting thrust. Taking into account the well-known formula of N.E. Zhukovskiy for the rotor thrust:

$$F = (33.25\,\eta_0\,DN)^{2/3},$$

where F – the rotor thrust;

D – the diameter of the rotor;

N – power on the motor shaft;

η_0 – efficiency factor of the rotor.

For the emergency balance of thrust, normalized by the value of the total thrust $F_{\Sigma} = n\ F_0$ until the moment of failure, taking into account the constancy η_0:

$$\frac{F_{rest}}{F_{\Sigma}} = \left(\frac{N_{rest}}{N_{\Sigma}}\right)^{2/3} = \left(\frac{n-1}{n}k\right)^{2/3}.$$

For the VTOL aircraft of the multipurpose cryogenic convertiplane type (MCCP) [13] (Fig. 3.5), with the number of PP engines 2, 3 and 4 respectively, the redistribution of thrust and PP engine power can be determined, with the maximum forcing speed of each engine k being equal to 1.1 taking into account the possibility of emergency forcing thrust by 10% (see Table 3.2).

Considering that

$$F_{\Sigma} = \varepsilon \cdot m_A \cdot g,$$

where ε – the thrust-weight ratio;

$m_A\, g$ – the weight of VTOL aircraft.

The emergency remaining thrust will be:

$$F_{rest} = \left(\frac{n-1}{n}k\right)^{2/3} \varepsilon \cdot m_A \cdot g.$$

Required stock of the engine, taken into account when choosing the engine ($\varepsilon \to \varepsilon_n$):

$$\varepsilon_n = \left(\frac{n}{k(n-1)}\right)^{2/3} \varepsilon.$$

At present, the practical use of VTOL aircraft for civil aviation is limited by the conflicting requirements for their PP: thrust generated in the modes of vertical takeoff, landing or hovering must be greater than its weight. At the same time, a large thrust-weight ratio from 1.12 to 1.36 in the total weight of the VTOL aircraft does not inevitably increase the weight share of the PP. This makes it possible to achieve an acceptable specific load on capacity and to ensure a significant increase in the service life. Therefore, for more economical cruising at subsonic speeds the relative thrust should be increased, allowing the performance of cruise flight with fuller use of the PP thrust and lower specific fuel consumption.

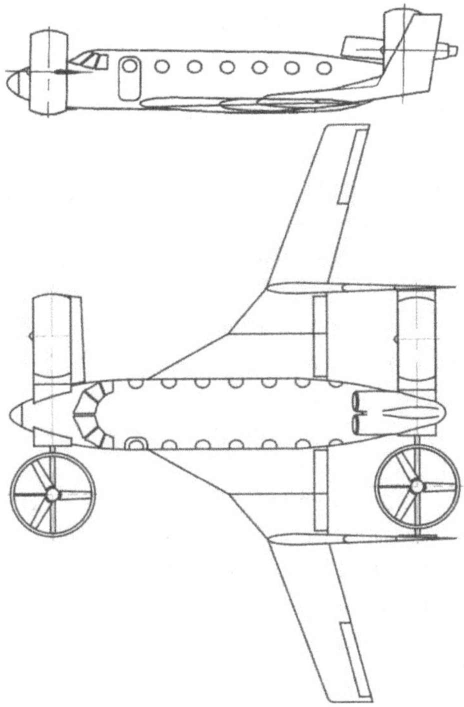

Fig. 3.5 The layout of the MCCP with twin-engine PP [13].

Table 3.2 Redistribution of thrust and engine power, according to the type of VTOL aircraft with synchronized rotor drive system in case of a single engine failure.

Number of engines, *n*	**2**		**3**		**4**	
The coefficient of emergency engine forcing, *k*	1.0	1.1	1.0	1.1	1.0	1.1
N_{rest}/N_{Σ}	0.50	0.55	0.67	0.73	0.75	0.83
F_{rest}/F_{Σ}	0.63	0.67	0.76	0.81	0.83	0.88
Thrust-weight ratio of VTOL aircraft ϵ	1.12	1.23	1.15	1.27	1.24	1.36
Required thrust-weight ratio ϵ_n	1.78	1.84	1.51	1.56	1.50	1.55

3.4 Problems of Stability and Controllability of Hydro Convertiplane with Tandem-Mounted Rotors in Rotary Annular Channels

When basing the HCP on water, it is necessary to ensure the protection of their engines and rotors from the ingress of sea water, as well as the stability of such a aircraft with a combination of its hydrodynamic, seaworthy and amphibious characteristics [15]. Horizontal HCP flight can occur in close proximity to the water surface or a flat piece of ground (screen) using the screen effect, which is formed at low altitudes due to the increase in lift and decrease in wing resistance [7].

The conducted studies [16, 17] of the conditions for ensuring the static and dynamic stability of the HCP with four tandem-mounted rotors in the rotary annular channels showed the following. The composite wing should have a developed center section, a slight elongation and a slight negative transverse *V*, as well as high-lying horizontal console. On consoles, which are practically unaffected by a screen, effective mechanization and front rotors in the rotary annular channels can be used. Horizontal tail should be sufficiently effective and taken out of the zone of influence of the screen.

Three-beam tail and automatic control systems with synchronous deflection of the front and rear (inter-beam) annular channels in opposite directions should ensure sufficient longitudinal stabilization and stability on the transition modes from screen to free flight and back.

Flight of the HCP near the screen is stable only in a limited range. To expand this range and when flying over a wavy screen, controls should be provided for stabilizing the HCP both in angle of attack and in height.

The most rational location of the center of gravity of the HCP with the front and rear rotors of the same diameter is the midpoint between their rotational annular channels. If the center of gravity between the rotors is shifted more than by 5–7% of the distance between the rotors, then unacceptable losses of vertical thrust in the hover mode appear, as well as an increase in the weight of the gearbox due to the unfavorable power distribution between the rotors [8].

The total area of the fins of the three-beam tail can be less due to the rotors in the annular canals, behind which an air jet is formed, which has high energy that makes it possible to install effective aerodynamic steering surfaces behind them in the course control mode [7].

Conventional three- and four-bladed rotors in rotary annular channels have high efficiency both in conditions of vertical takeoff, landing and hovering (up to 85%), and in cruising flight at subsonic speeds (up to 80–87%) [8]. Thus, if rotor sizes in the rotary annular channel and the load on rotor disk area are not limited by the HCP construction, the high levels of thrust are reached as under static conditions (at hover, for example), and in cruising flight conditions.

The tandem arrangement of the rotors in the rotary annular channels meets the requirements for providing the necessary control points in the hanging mode without using a complex rotor with a cyclic pitch change or auxiliary control systems (antitorque tail rotor). Longitudinal controllability is provided by means of differentiated changes in the angle of installation of the front and rear rotors. The transversal control is provided by increasing the angle of installation of the blade for the front and rear rotors on one side of the HCP and for reducing the angles of installation of the blade for the front and rear rotors on the other. The system of increasing the stability of the HCP in pitch and roll should exclude the influence of the emerging instability along one axis on the stability along other axes [7].

The rotors located along the diagonal have the same direction of rotation and the yaw moment can be reached by small deviations of the axes of the rotors. For example, if the angle of deviation is measured from the direction of flight, then the angle of rotation of the front rotors can be equal to 94°, and the rear rotors to 85°. As shown in Fig. 3.6, the yaw moment M_y

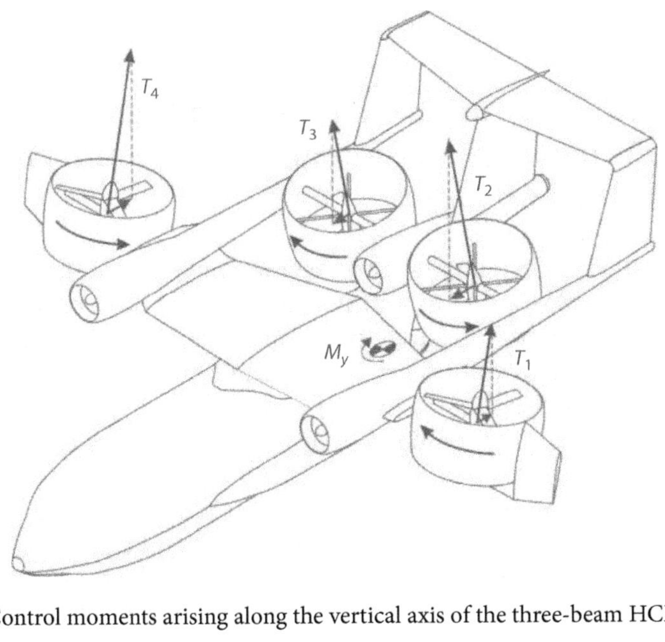

Fig. 3.6 Control moments arising along the vertical axis of the three-beam HCP.

occurs if the blade installation angle and, therefore, the power increases on two rotors and simultaneously decreases on the other two rotors [9]. The full yaw moment is formed as a result of the interaction of the horizontal rotor thrust components that create the unfolding moment.

With this system, track control along the course is provided without changing the pitch, roll or vertical thrust. A yaw control system of the HCP with tandem-mounted rotors can reduce the vertical thrust by about 1%, but it does not degrade the acceptable characteristics of its stability and controllability. And if additional power is required to control the course, this problem can be solved by increasing the angles of rotation of the rotors [9].

Considered as a whole, the tandem arrangement of the rotors in the rotary annular canals has good flight characteristics and handling characteristics required for an aircraft used as a helicopter, a plane, and a wing-in-ground-effect-craft.

In order to reach the best characteristics of the HCP at the vertical takeoff and during a transient flight mode, it is necessary to make maximum use of the available power of the engines, taking into account the possibility of failure of one of them. The required number of engines, taking into account the failure of one of them with a total thrust-weight ratio of 1.1 to 1.3, should be three [8]. This means that it is necessary to select such HCP, the failure of one of them will only reduce the propulsion of the powerplant without loss of controllability and a noticeable decrease in the load on the rotor disk area and on the wing, which should be significant for high-speed HCP.

3.5 Cryogenic Turboelectric Aircrafts are a Good Solution for Short-Range and Takeoff Hybrid Airline Complexes

The experience of operating jet passenger aircraft showed that the volume of air traffic increases due to an increase in traffic on local airlines. But neither purely freight nor purely passenger regional convertiplanes with takeoff and landing within the city limits can be used profitably. To solve this problem, it is necessary to use a special transport system for local airlines, created on the basis of high-speed MCCP of the vertical and landing (VTOL) and short takeoff and landing (STOL) [18, 19].

According to experts' estimates [20], the MCCP on liquefied natural gas will, on average, allow a 2.5-fold decrease in fuel costs and a reduction in direct operating costs by 4–5% per flight hour. This increases the reliability and stability of control in transient flight regimes, which allows to improve the corkrotor characteristics and reduce the loss of aerodynamic quality for MCCP balancing by up to 20%. The anterior horizontal tail works in conjunction with the wing and creates additional lift and unloads the wing [13].

To eliminate the known drawbacks of jet passenger aircraft in Europe and the United States, electric and aircraft construction companies have joined forces and are already conducting research on new generations of electric airplanes based on electric motors powered by fuel cells based on hybrid lithium-ion batteries having a fast (within half an hour) recharge time. Such electric airplanes of the new generation will be noticeably quieter than traditional analogs and have the ability to perform STOL [21].

Extensive operational requirements for hybrid aircraft of the new generation will lead to the development of a number of models of unmanned and cryogenic turboelectric aircrafts. The creation of such aircrafts using parallel-serial hybrid technology with a gas turbine engine (GTE), working on jet fuel and liquefied natural gas, and an electric motor-generator will double the horizontal thrust-to-weight ratio [22].

A feature of the use of parallel-serial hybrid power drive technology in a cryogenic turbo-electric aircraft is the scalability, which allows to create light medium-range aircraft with a seating capacity of 30 and 48 people [18]. Moreover, such an aircraft with a total fuel efficiency of 10.6 g/pass·km can have three electric motors with a peak power of 1,400 kW (nominal 770 kW) each and a cryogenic gas turbine engine for liquefied natural gas of the type PK-65VF (power 1100 hp). Creation of a more powerful PP using parallel-serial hybrid technology with three peak electric motors with a rated power of 4540/2500 kW each and a cryogenic gas turbine engine for liquefied natural gas (with a capacity of 2,800 hp) allows to master a medium-range aircraft for 110 passengers.

Liquefied natural gas will improve operational performance and indicators of transport and environmental efficiency, including achieving (when performing the STOL) fuel efficiency of 10.8-11.2 g/pass·km [20].

The development of conceptual approaches requires the promotion of innovative aircraft development projects based on technical solutions [4, 12, 23]. Their implementation will allow the mastering of a number of models of light and medium convertible high-speed helicopters with turbo diesel and gas turbine engines in the PP. Innovative aircraft of this kind can perform a vertical or short takeoff/landing (Fig. 3.7). Mastering of such aircrafts will allow the vertical takeoff at normal takeoff load and short takeoff with an overload of 6%–13% more of the normal takeoff weight, while ensuring the required flight range.

To eliminate or reduce the impact of the shortcomings inherent in a number of recently developed twin-rotor longitudinal helicopters with a

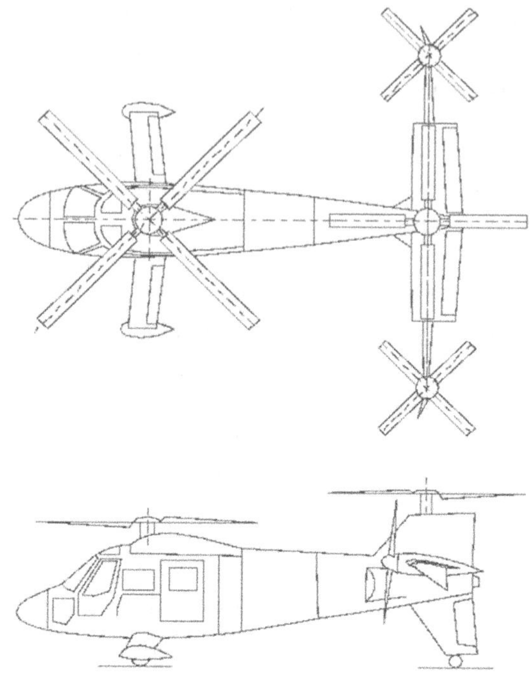

Fig. 3.7 Aerodynamic scheme of light and medium convertible speed helicopters [4].

pushing rotor, diesel PP can be used arranged on a parallel-serial hybrid technology [24], that will allow a cruise flight for 4 or 6 hours, respectively, when performing a vertical or short takeoff/landing [4].

On the basis of a helicopter with two longitudinal rotors it is possible to master a diesel-electric lightweight convertible high-speed helicopter with a takeoff weight of 900 and 1,040 kg and to carry 0.2 and 0.3 tons of cargo at a range of up to 1,760 and 2,800 km, respectively, when performing vertical and short takeoff/landing. The increase in flight time and generating capacity for electric supply is achieved by using the energy of the incoming flow, due to a multimode hybrid PP with reversible electric motors generators. During cruising flight, they receive energy from the incoming air flow in the course of electric wind generators; they rotate from two autograting main rotors. This should allow the achievement of flight time up to 3 hours when performing vertical and 5 hours when performing short takeoff/landing.

This parallel-serial hybrid PP has front and aft reversible electromotors-generators for power supply and two electric motors with a total peak/rated power of 50/28 and 22/12 kW and 50/23 kW, respectively. It is equipped with generator turbo diesel engines, which can provide another 66 and 44 kW to the main rotors of the longitudinal group.

A battery weighing about 224 kg (33% of the mass of an empty aircraft) will allow a short takeoff and landing to reach a flight range of 390 km at a cruising speed of 385 km/h. When the battery charge drops to 25% of the maximum value, the turbodiesel engine will turn on and, rotating the reversible front electric motor-generator, in flight will feed the accumulator along with the aft reversible electric motor-generators, the rotating main rotor from the incident flow.

According to the forecasts of specialists [6, 20, 25, 26], not only high-speed and combined helicopters, but also transformed rotary-wing aircraft of the new generation will have the greatest value in the converted helicopter aircraft that are airplane-like convertible rotor aircrafts and convertiplanes. The foregoing suggests the real possibility of mastering the heavy convertible rotor aircrafts and convertiplanes (HCVP) of the new generation in the near future (Fig. 3.8).

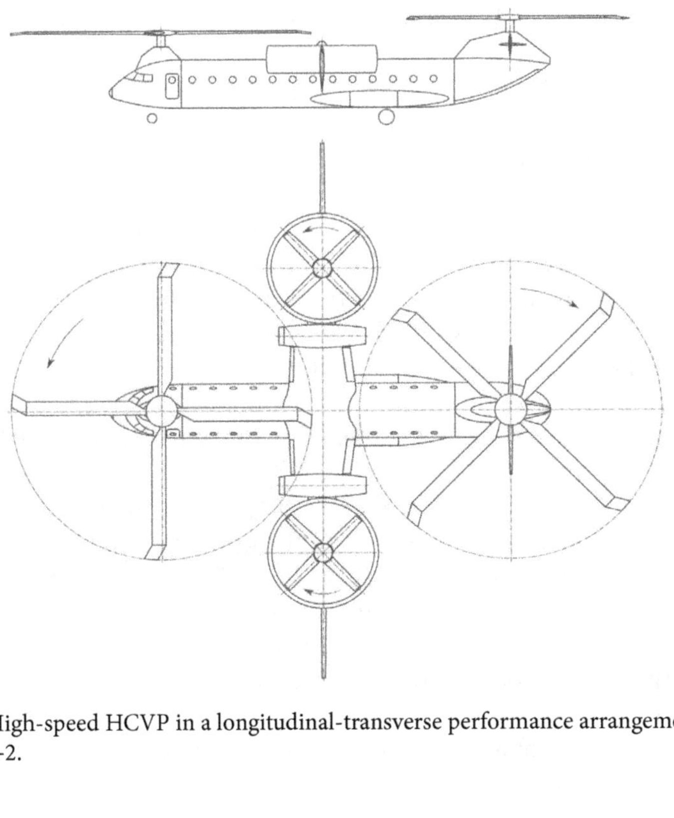

Fig. 3.8 High-speed HCVP in a longitudinal-transverse performance arrangement of rotors X2+2.

The most relevant in modern conditions may be the priority mastering of HCVP-80 (Type 1.1) with D-12V type engines (N = 5,500×2) and rotors in cantilever rotary annular canals (see Table 3.3). Since larger main rotors, for example, HCVP-80 in its flight configuration of a helicopter and rotor aircraft, are designed to create lift, the translational motion in horizontal flight is largely provided by cantilever rotors in rotary annular channels.

On helicopter flight modes, such a four-rotor main carrier with a multimode aerodynamic control system and cantilever main rotors providing additional lift force allows for the implementation of the technology of VTOL and STOL, while this aircraft has 6–7% less fuselage length with rotors and has less overall dimensions when VTOL is fulfilled.

This will save space for the takeoff site; therefore such HCVP versions with larger main rotors and cantilever rotors in the rotary annular channels have a minimum takeoff platform and the maximum ratio of payload to the length of the takeoff platform when performing VTOL.

Extensive operational requirements for convertible rotary-wing aircrafts of the new generation will lead to the creation of HCVP that can compete with high-speed helicopters, as well as provide three specific flight modes: helicopter, rotary-wing and gyroplane. The last two modes make it possible to reach cruising speeds on these safe flight modes of 550 and 620 km/h, respectively.

3.6 Conclusion

Aircraft of the traditional aerodynamic layout need large aerodromes that should be carried far beyond the city limits. And the time required for a trip from the city to the aerodrome and from the aerodrome to the city can be commensurate or even longer than the flight time itself. The task of creating convertible helicopters-airplanes that could carry out a vertical takeoff and landing on limited-size sites within the city boundaries, becomes urgent.

A feature of convertible helicopters is the management of vertical takeoff and landing modes and transient modes, when the aerodynamic control surfaces are inefficient. Their power plant must provide vertical thrust, exceeding in size the takeoff weight of the aircraft. This requires the creation of more economical and sophisticated gas turbine engines. The traditional aerodynamic scheme of such aircraft, in which the main lift force is created by the wing, being the main aerodynamic carrier surface, has a rather large specific load on the wing (about 460–525 kg/m^2), which will increase in proportion to the increase in size. This is confirmed by comparison with the traditional schemes of turboprop aircraft.

Table 3.3 Preliminary tactical and technical characteristics for multiple-rotor HCVP.

No.	Parameters	Values		
		Type 1.1	Type 1.2	Type 1.3
1.	Sizes according to the arrangement type:	HCVP-80	HCVP-120	HCVP-165
1.1	length with main rotors/fuselage, m	36.56 / 22.3	44.86 / 27.4	52.0 / 32.86
1.2	total height on chassis at VTOL, m	6.5	7.5	8.1
1.3	wing span with annular channels, m	18.0	22.0	23.67
1.4	wing area with annular channels, m^2	54.0	80.66	93.34
1.5	folded wing span, m	9.8	11.8	12.7
1.6	*min* takeoff area with VTOL, m^2	658	986	1230
1.7	parking ramp, m^2	218	323	417
1.8	specific takeoff opportunity – relation of paragraph 3.3/ paragraph 1.5 at VTOL, kg/m^2	12.16	12,19	13.41
2.	The powerplant with two turboshaft engines of "Motor-SICH" OJSC, model	D-12V	D-136	D-27
2.1	total power, hp	7,500 × 2	11,400 × 2	14,000 × 2
2.2	thrust-weight ratio under paragraph 3.1, kgf/kg	1.44	1.44	1.44

(*Continued*)

Table 3.3 Preliminary tactical and technical characteristics for multiple-rotor HCVP. (*Continued*)

No.	Parameters	Values		
		Type 1.1	**Type 1.2**	**Type 1.3**
3.	Masses and loads (power on main rotor):	(11,250)	(18,240)	(21,000)
3.1	normal at the vertical takeoff, kg	33,000	50,160	61,600
3.2	at the short takeoff, kg	38,700	59,760	72,800
3.3	normal payload at the takeoff under para. 3.1/para. 3.2, people (t)	3 + 80 (8.0)/ 3 + 96 (12)	3 + 120 (12)/ 3 + 130 (19)	3 + 162 (16.5)/ 3 + 162 (25)
3.5	empty kg	19,500	29,960	36,800
3.6	fuel supply according to para. 3.1/para. 3.2, t	5.2 / 6.9	7.9 / 10.5	8.0 / 10.7
4.	Rotor diameter:			
4.1	diameter of larger main rotors (D), m	17.2 × 2	21.21 × 2	23.52 × 2
4.2	diameter of smaller rotors (d) in rotary annular channels, m	4.5 × 2	5.5 × 2	6.1 × 2
4.3	Rotor disk area, m^2	496.25	753.77	926.92
4.4	The rotation speed of the smaller rotors: at the vertical takeoff, min^{-1} during cruising flight, min^{-1}	1,225 980	1,000 800	900 720
5.	Criterion para. 9.3 at the VTOL, t·km	13,500	21,600	29,700

(*Continued*)

Table 3.3 Preliminary tactical and technical characteristics for multiple-rotor HCVP. (*Continued*)

No.	Parameters	Values		
		Type 1.1	Type 1.2	Type 1.3
6.	The specific load on the rotor disk area under para. 3.1, kg/m^2	66.5	66.5	66.46
7.	Specific load on power, kg/hp	2.2	2.2	2.2
8.	The specific load on the wing with takeoff weight under para. 3.2 kg/m^2	716.7	740.9	780.0
9.	Flight performance:			
9.1	cruising speed at 25% of the available power of PP, km / h	620	580	550
9.2	maximum speed, km/h	650	610	580
9.3	flight time under para. 3.1/para. 3.2, h	4.0/6.0	4.0/6.0	4.0/6.0
9.4	the length of the flight under para. 3.1/para. 3.2, km	2,560/3,720	2,320/3,480	2,200/3,300
9.5	service ceiling, m	8,300	7,900	7,600
9.6	duration of hover, h	0.6	0.6	0.6
9.7	takeoff distance when landing at the ground roll/short take-off, m	135/90	150/100	165/110

The fleet of cryogenic turbo-electric airplanes on liquefied natural gas will improve operational performance, transport and environmental performance, as well as achieve (when performing STOL) fuel efficiency of 10.8-11.2 g/pass·km.

Light and medium models make up more than 80% of the entire helicopter fleet; therefore, local convertible air carriers will be interested in medium-sized convertible vertical or short takeoff and landing aircrafts, which can carry at least 24 passengers with a cruising speed of 500-650 km/h and flight altitude up to 8,000 m on routes up to 1,300 km long with take-off and landing within the city.

References

1. Thornborough, A.M., *Bell-Boeing V-22 Osprey: An Aeroguide Special*, A.M. Thornborough (Ed.), p. 40, Linewrights Ltd. P O Box 832, Ongar, Essex CM5 0NH, England, 1990.
2. AW609 Tiltrotor Program Update. AgustaWestland - Finmeccanica: [website], URL:http://www.agustawestland.com/news/aw609-tiltrotor-programmeupdate.
3. Khurana, K.C., *Aviation Management: Global Perspectives*, K.C. Khurana (Ed.), p. 316, Global India Publications, New Delhi, 2009.
4. D.S. Durov, Easy convertible speed helicopter. RF Patent No. 2579235S1 IPC B64C 64/22 (2006.01), publ. 04/10/2016 bul. No. 10.
5. Durov, D.S., Hydro convertiplane – new opportunities for modern aviation. In the collection of: Scientific articles *The study of the performance and reliability of mechanical systems*. Hands dep. VINITI number 1415-B2006 of September 13, 2006.
6. Ruzhitsky, E.I., *American vertical take-off aircraft*, 192 p, Astral AST, M, 2000.
7. D.S. Durov, *Hydro convertiplane*, 15 p, Patent of the Russian Federation 2264951C1. IPC B 64 C 35/00, 29/00, 37/00, B 60 V 1/08. Publ. November 27, 2005 Byul. No. 33.
8. Pavlenko, V.F., *Vertical take-off and landing*, 112 p, Voenizdat, M., 1968.
9. Pavlenko, V.F., *Aircraft vertical takeoff and landing*, 160 p, Military Publishing, M, 1965.
10. Donald, D., *Modern military aviation and air forces of the world*, 192 p, ZAO Omega, M, 2003.
11. Hafer, K. and Sachs, G., *The technique of vertical take-off and landing*, 376 p, Mir, M, 1985.
12. Durov, D.S., Multi-screw high-speed aircraft-like rotary-winged aircraft - prospects for regional convertible rotary-wing aircraft. Modeling and analysis of complex technical and technological systems. *Sat. articles of the International Scientific and Practical Conference (November 1, 2017, Volgograd)*, AETERNA, Ufa, pp. 33–38, 2017.
13. D.S. Durov, Multipurpose cryogenic convertiplane, 15 p. The patent of the Russian Federation 2394723C1, MPK B 64 C 35/00, 29/00, 37/00, B 60 V 1/08. Publ. 07/20/2010, Bull. No. 20.

14. Durov, D.S., Features of the management of the vertical take-off and landing aircraft in the transitional and hovering mode. *"Practice and prospects for the development of partnership in the field of higher education": Materials of the 14th international scientific-practical seminar*, vol. 3, DonNTU, Donetsk, pp. 95–103, 2013.
15. Bratukhin, A.G., Valuev, N.O., Gilberg, L.A., and others., *Naval Aviation of Russia*, 240 p, Mashinostroenie, M, 1996.
16. Durov, D.S., Problems of stability and controllability of hydroconvertocrano-plane with tandem screws in rotary annular channels. *Materials of the 8th International n-practical. seminar "Practice and prospects of development of partnership in the field of higher education"*, book 3, Donetsk, Taganrog, pp. 183–194, 2007.
17. Shuvalov, A., *Ekranoplan - Problems and Prospects*, pp. 22–24, General Aviation, No. 4, 2003.
18. Durov, D.S., Convertible with cryogenic power plants is the priority of rotary-wing business and carrier-based aircraft. *Materials of the 12th International practical of the seminar "Practice and prospects of development of partnership in the field of higher education"*, vol. 1, Publishing house Don-NTU, Donetsk - Taganrog, 250 p, 2011.
19. Durov, D.S., The use of convertible rotary-wing aircraft in the function of expanding the capabilities of transport aircraft. *"Practice and prospects for the development of partnership in the field of higher education": Proceedings of the 14th international scientific-practical seminar*, vol. 3, DonNTU, Donetsk, pp. 104–109, 2013.
20. Yeger, S.M., *Fundamentals of aviation technology*, 720 p, Mechanical engineering, M, 2003.
21. Bowers, P., *Aircraft unconventional schemes*, 320 p, Mir, Trans. from English M, 1991.
22. D.S. Durov, *A cryogenic turbo-electric aircraft of short takeoff and landing*, 14 p, The patent of the Russian Federation 2534676C1. Publ. 12/10/2014 Bull. No. 34.
23. Durov, D.S., Prospective multipurpose sea-and ship-based helicopters. Modeling and analysis of complex technical and technological systems. *Sat. articles of the International Scientific and Practical Conference (November 1, 2017, Volgograd)*, AETERNA, Ufa, pp. 29–33, 2017.
24. Durov, D.S., *Hydro convertiplane*, 14 p, The patent of the Russian Federation 2351506C2. IPC B64C 37/00 (2006.01), B64C 35/00 (2006.01), B64C 29/00 (2006.01). Publ. April 23, 2007 Byul. No. 10.
25. Cabrit, P., Fast rocketcraft LifeRCraft IADPD. Clean Sky 2. Information Day dedicated to the 1st Call for Proposals, 2015.
26. Mikheev, V.R., *The development of rotorcraft circuits*, 240 p, Mashin, M, 1993.

4

Conceptual Design of A Multifunctional Amphibious Plane

V'iacheslav V. Orekhov

Southern Federal University, Engineering Technological Academy, Department of Engineering Graphics and Computer Design, Taganrog, Russia

Abstract

The fourth chapter introduces the conceptual model of a multifunctional amphibious plane. The brief technical characteristics, visual imagery and historical aspects of the plane creation are given. Rendering of the future model is given. Consolidated PBY-5A Catalina, a modification of an amphibious plane of the 1950s, became the prototype of the amphibious plane concept. In addition to use for travel and outdoor activities, this plane can be applied in many other utilitarian purposes.

Keywords: Amphibious plane, conceptual design, prototype, computer simulation, rendering

4.1 Introduction, Historical Stages

Conceptual modeling is applied in all areas of industrial design today. This paper presents a conceptual model for a multifunctional amphibious plane. The historical review of the "Consolidated PBY Catalina" plane, which served as the prototype for the development of the design of a new multifunctional amphibious plane (Fig. 4.1), is presented.

At the beginning of the 1930s, the American firm Consolidated Air-Craft began developing a new patrol bomber to intercept sea convoys in the event of war in the Pacific, to the order of militaries. The work was led by engineer Isaac Laddon. According to the technical requirements of the

---

Email: orekhovich@yandex.ru

Iftikhar B. Abbasov (ed.) Computer Modeling in the Aerospace Industry, (161–178)

Fig. 4.1 Amphibious plane Consolidated PBY Catalina, a prototype of the developed model.

new patrol flying boat, it was to significantly expand the fleet's ability to conduct long-range maritime reconnaissance. The plane essentially had to be a flying boat and an amphibian at the same time. Equipped with two engines, an all-metal monoplane-parasol, in aerodynamic terms the plane was a monoplane with a wing raised above the fuselage-boat on a voluminous pylon [1–3]. The calculated characteristics of the model in many of their parameters exceeded the task. Tests showed that the flight data of the new hydroplane exceed the most optimistic expectations: the flight range reached 6,900 km, the maximum speed - 275 km/h, the service ceiling - 5,760 m.

The plane was built in the United States at the plants Consolidated (San Diego, New Orleans) and NAF (Philadelphia), in Canada at Boeing (Vancouver) and Canadian Vickers (since 1944 – Canadair, Montreal), and in the Soviet Union. In October 1936, deliveries of the Catalina production planes began. Production continued until 1945; the firm Consolidated produced 2,394 planes. In the USSR, the factory number 31 in Taganrog, where 27 units were produced became the place of production. The processing of the drawings and other technical documentation, as well as the management of the launch into the series, were entrusted to the experimental design office of Georgiy Mikhailovich Beriev [4, 5].

In February 1937, to familiarize themselves with the production and acceptance of documentation and equipment, a group of plant specialists and design bureaus were sent to the plant in San Diego (USA), which produced a PBY-1 hydroplane. Part of the planes transferred to civil aviation were named MP-7. The USSR received three planes "Catalina" as the samples used for the licensed production. In May 1938, the first submarine with the American Wright Cyclone GR-1820-G3 engines was built. Factory tests began on June 3; the plane had good stability on all operational flight modes, which allowed it to quickly master it. In January 1939, the serial construction of a flying boat under the designation THP (Transport Hydroplane) began (Fig. 4.2) [6, 20].

Unlike the American prototype, the modifications of the THP were equipped with various engines of domestic production. The plane was a twin-engine flying boat of metal construction – a strut-off monoplane with a high wing. The wing was installed above the fuselage on a streamlined pylon. The hull of the boat was relatively wide, low and rounded at the top. The fuselage was divided into five duralumin waterproof bulkheads with hermetic hatches for passage into six isolated compartments. This ensured buoyancy in case of damage to the cladding of the bottom of one or even several compartments.

In the bow there were hatches for the anchor and mooring cable. The crew of the hydroplane consisted of two pilots, navigating officer, radio operator, flight mechanic and two gunners. The plane had a rather complicated and time-consuming design. For this reason, as well as in connection with the organization of the mass production of the Che-2 (MDR-6) flying boat of I.V. Chetverikov (Fig. 4.3), the production of the THP was discontinued at the end of 1940 [6, 20].

The Catalina plane was recognized by the Soviet Union as better than its counterparts created by its own designers, and built under license during the war. Moreover, by the end of the war, Catalina was produced in many new versions and sold better than new designs. The number of built Catalinas was greater than any other flying-boats or hydroplanes in history.

Airplanes exported to the UK, from the end of February 1940, were refined by the Cairds Yard (Greenock), Scottish Aviation (Prestwick) and

Fig. 4.2 THP Hydroplane.

Fig. 4.3 Hydroplane of I. V. Chetverikov Che-2 (MDR-6).

Saunders Row (Bomeris) plants. During the revision, radars, impulse searchlights and instruments of English production were installed, redan was reinforced. English hydroplanes "Catalina" began to be used in the Atlantic in March 1941. On April 29, the Catalina Second World War flying boat sank its first enemy submarine. In May, they participated in the hunt for the German battleship "Bismarck", directing ships and planes to it. At the same time, they were first attracted to escorting convoys from America [2, 4].

In 1942–43, the British Air Force used flying boats of this type in the northern and southern parts of the Atlantic, in the Mediterranean, Indian Ocean and Arctic waters as long-distance reconnaissance planes. The planes carried out rescue work. In Southeast Asia, hydroplanes were used to put ground mines. Since the autumn of 1939, American planes began systematically patrolling the 200-mile zone along the Atlantic coast of the USA, and then practically the entire shipping route to England. In the Pacific, the Black Cat squadron became famous–they used the Catalina for night strikes on Japanese ships.

Canadian planes basically provided protection from submarines in the North Atlantic. Dutch planes 28-5MN from January to March 1942 were fighting against the Japanese, then were evacuated to Australia and Ceylon. Australian and New Zealand cars fought in the Pacific before the defeat of Japan.

The plane was removed from production in the USSR in November 1940, in Canada in May 1945, in the USA in September 1945. These machines were taken off in Canada in the autumn of 1945, in the UK at the end of 1946, in Australia in 1953, and in New Zealand in December 1956. In the USA they flew in the fleet aviation until 1949 (they were in reserve until January 1957), in the Air Force until January 1953, and in the Coast Guard until April 1954.

In the Soviet Union, the last hydroplane of this type was operated in the Northern Fleet until the autumn of 1945. In polar aviation, they served until mid-1948; the civilian one, the MP-7, flew until 1949. Since 1952, the flying boats of the Second World War of American manufacture began to be replaced by the Be-6 hydroplanes (Fig. 4.4). They flew in the Baltic Fleet until May 1954, on the Black Sea until October 1955, and in the Pacific until August 1957. In civil aviation, the last cars in the Far East were written off in 1958.

After the end of World War II, the Catalina amphibious planes were put into service in Argentina, Brazil, Denmark, the Dominican Republic, Israel, Colombia, Indonesia, Mexico, the Netherlands, Norway, Peru, France, Sweden, and Japan. The Dutch cars participated in the colonial war in Indonesia in 1946–1948, the Israeli cars in the war with Egypt in 1956. The last combat operation for the Catalina hydroplanes was an operation in Cochinos Bay in April 1961.The last combat operation for the Catalina hydroplanes was an operation in Cochinos Bay in April 1961.

The enormous popularity of the plane among the military during the war years was due in large part to its simplicity and reliability. This is what causes the same popularity of a hydroplane among civilian users. Numerous modifications of the plane made it indispensable in peacetime. Thus, the disarmed plane, capable of taking off from any suitable reservoir, delivered people and goods, for example, to Alaska and to the Amazon delta. The recipe for success in a civilian "old women" hydroplane, is simple: dismantle the entire military arsenal, install passenger seats and now it is a universal passenger liner, which is time-tested. It can perform passenger and freight transport, meteorology, fire patrols, fire extinguishing, a mobile research laboratory and many peaceful tasks.

The first civilian PBY was sold in 1937 to the American Museum of Natural History for an expedition to New Guinea by Dr. Erchbold and was given the name GUBA (the Papuan "severe storm" serial number - NC 777). But it was bought up by the Soviet government for a large-scale operation to search for the crew of Levanevsky, who disappeared during

Fig. 4.4 Be-6 Hydroplane.

a transpolar flight (Fig. 4.5). PBY-4 is a modification, the plane of which were received by blisters on onboard rifle installations, which later became the "brand name" of "Catalina".

After the war, a large number of Catalinas became available in the market of decommissioned military equipment, where they were willingly bought by operators of charter and cargo airlines. After the dismantling of weapons, reservations and parts of the radio equipment, the plane's carrying capacity increased significantly. In addition, retired patrol cars could be used with improvised runways and bodies of suitable size, which was a significant plus for their new owners. The "Catalinas" were widely used for passenger traffic in the Caribbean and Alaska, but their main task was still the delivery of goods and people to remote and hard-to-reach places.

The Consolidated PBY-6A Catalina flying boat acted as a research plane for the team of researchers led by Jacques-Yves Cousteau (Fig. 4.6). In 1967,

Fig. 4.5 Amphibious plane PBY Catalina with tail number NC 777.

Fig. 4.6 Research amphibious plane of Jacques-Yves Cousteau's team.

Cousteau along with his father became a member of the famous expedition Calypso, whose goal was to film sharks living in the Red Sea and the Indian Ocean.

4.2 Concept

The design of the Catalina allows you to use it in almost any conditions, both with and without an equipped airfield, using the surface of a reservoir. In 1950, one of the modifications of the Consolidated PBY-5A Catalina was a flying yacht (Fig. 4.7)–a flying yacht or an amphibious plane, depending on preferences.

Inventor Glen Odekirk saw in these machines a new future for his idea. He thought of airplanes as the embodiment of chic, glamour, luxury and elitism. "An airplane is just a vehicle in which you move from point A to point B.," Odekirk said in an interview in 1953, "the difference is only in the level of comfort. Therefore, I decided to create a truly unique plane in which a person could even live in comfort." The main innovation of

Fig. 4.7 Advertising photo in LIFE journal, with an elite flying yacht.

his flying amphibious plane was high noise isolation. The 4-inch thick fiberglass needed $265,000 at the rate of those years. Today it is an even more impressive amount. This price did not include installation and maintenance. And in combination with other operating costs, this flying luxury became unattainable even for millionaires. There were eight sleeping berths: three double beds and two single beds. Next to each bed there was a personal light, radio, air conditioning hole and a telephone. In addition, a TV was built into the plane. On board there was a shower with hot and cold water, and a toilet, a small galley with white porcelain and stainless steel, a stove, an oven, a fridge and a freezer.

It was the idea of a flying yacht that formed the basis for the development of the design concept of a multipurpose amphibious plane. It should be a flying yacht that can accommodate a small group of people for an active holiday and has all the necessary amenities on board [7, 8]. In addition to the use in travelling and active rest, this plane can be applied in other purposes, including passenger transportation on local hard accessible roads, ambulance vehicles and transport for medical personnel as a mobile medical office, fire extinguishing and territory patrol, search and rescue operations.

The design of the model being developed is similar to the Catalina: a monoplane with a wing raised above the fuselage-boat on a voluminous pylon in which the exit to the upper deck is located. The tail is single-fin. The main wing of a trapezoidal shape. The wing is equipped with a mechanization system, the deck above the main building is made in the form of wooden flooring [9, 10]. Exit to the deck is through the central pylon from the lower cabin. There is also an access to the main wing from the deck. The powerplant is represented by two turbo-screw engines. To get to the ground from the water, the plane is equipped with a landing gear at three points.

The development of the design concept involves the elaboration of cabin layout schemes. This article discusses the passenger cabin layout (Fig. 4.8) with four rows of seats with a center passage, this layout accommodates 28-30 passengers [11, 12].

The cabin has a toilet and luggage compartment, access to which is opened through the rear hatch. The cabin height is assumed to be 2 m, which will allow passengers to move freely around the cabin (Fig. 4.9). The rows of seats are placed on the right and left sides on the podiums [13].

As a result of a sketchy search for the image of the future vessel, a number of solutions were found [10, 14]. The volumetric-spatial solution is presented in sketches (Figs. 4.10, 4.11). Compared with the prototype, the projected model has become larger. Keeping proportions,

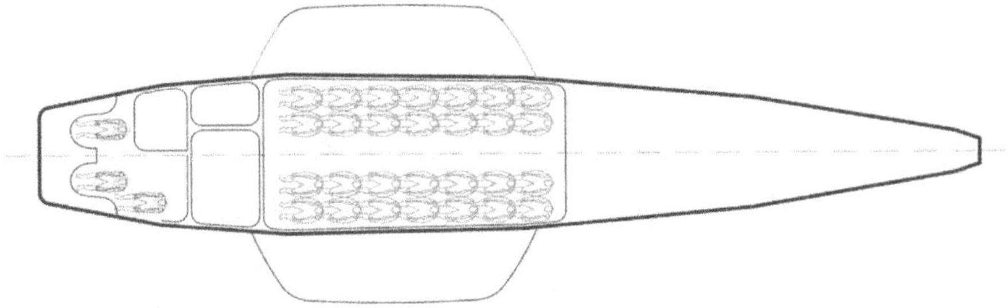

Fig. 4.8 Layout diagram of a passenger modified hydroplane.

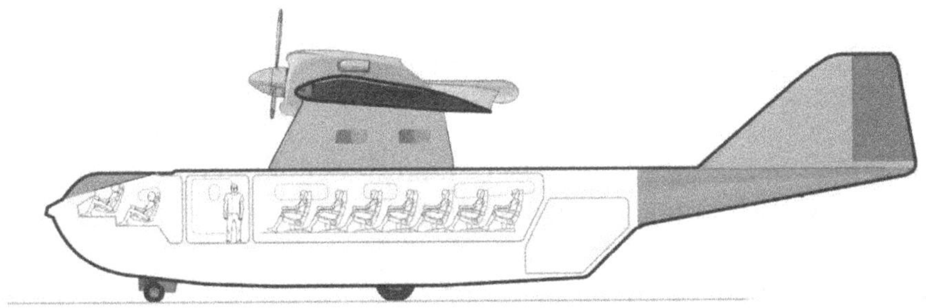

Fig. 4.9 Profile drawing of the plane cabin.

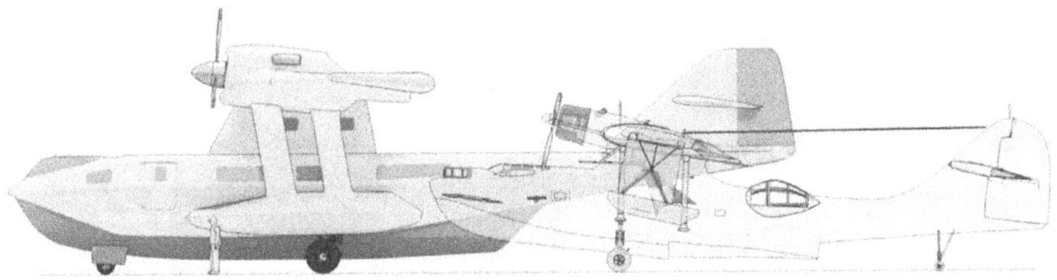

Fig. 4.10 The proportions of the new model and prototype. Side view.

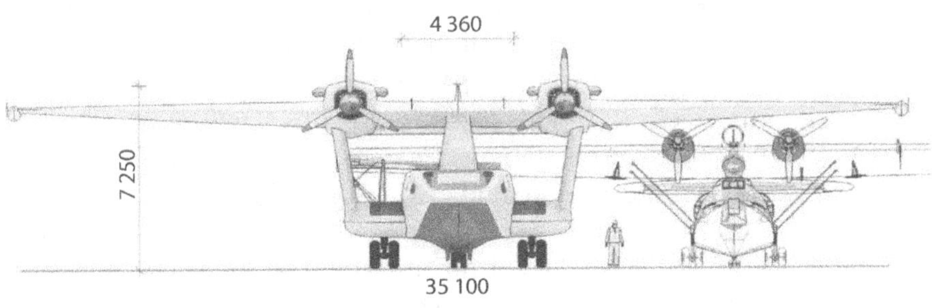

Fig. 4.11 The proportions of the new model and prototype. Front view.

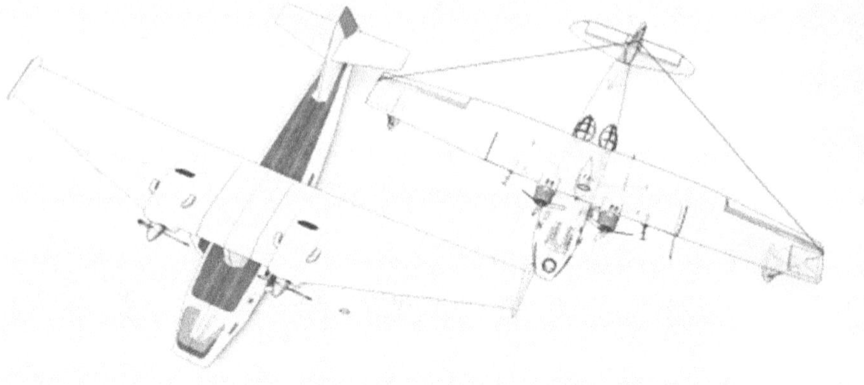

Fig. 4.12 The proportions of the new model and prototype. General view.

the dimensions of the model were increased. Preserving proportions allowed to stick to the external silhouette as the prototype of the concept model.

The increase in size is aimed to accommodate a comfortable cabin, and everything necessary for passenger comfort [15]. The length of the fuselage is 25.4 m; wingspan is 35.1 m. Height from ground level to the top of the tail unit is 7.25 m. A general view of the new model in comparison with the prototype is shown on the picture (Fig. 4.12).

4.3 3D Modeling

Along with many 3D modeling packages, 3D Studio Max allows the development of projects of almost any complexity [16]. The 3D Studio Max graphics system allows working with drawings made in other graphics packages, thereby providing the user with a lot of work space [17]. To create a three-dimensional model, there are a number of methods. To begin work, schematic projections are required [18] of the object being modeled. The modeling process begins with the creation of three perpendicular planes, with images of projections placed on them.

Next, a set of fuselage sections is created. These sections are placed along the axis of symmetry. A spatial grid is formed on the basis of these sections. The next step is to create a surface based on the created grid. The main fuselage lines are based on the Edit Poly polygon modeling. Extrude modifier is a gradual increment of the model along one or another coordinate axis.

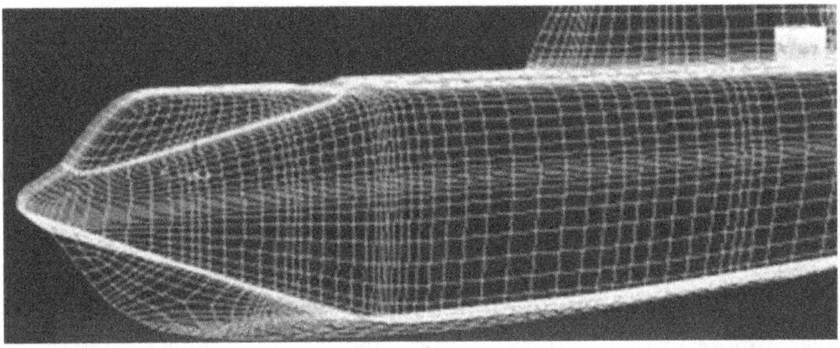

Fig. 4.13 Grid density in wireframe view mode.

The software package 3D Studio Max along with other methods allows the creation of models using High-Poly modeling. In order for the edges of the model not to have a faceted appearance, it is necessary that the polygons be small and the surface of the object consists of small planes; it is possible to observe this in the Wireframe grid view mode (Fig. 4.13).

To create a realistic and reliable model of the body in the program 3D Studio Max, plans and sectional views of the future body can be used. The standard scheme of High-Poly modeling occurs with a gradual increase in the level of detail of the 3d object:

- the first level is basic, and represents the overall shape of the object;
- at the second level, the base shape is refined, usually by adding chamfers;
- the third level is the final, it produces a clear detail of the object, usually through the use of smoothing plug-ins.

Then the case model is corrected and refined, and the TurboSmooth modifier is applied at the final stage. This algorithm is applied to all components of the plane: the fuselage, center section, tail, and side floats. To speed up and simplify the work, the model is simulated and assembled from one of the halves along the longitudinal axis of symmetry. The mirror half is created using the Mirror reflection method.

4.4 Application of Materials, Rendering

The next step is to configure the model for the application of materials and create them. The V-Ray visualizer from ChaosGroup was used to create the rendering. This software package contains a lot of parameters that allow control of the quality of the scene rendering. The visualizer requires the use and configuration of its VRayMtl shaders (materials) other than the standard ones. VRayMtl is the main, most used and versatile material from all offered by ChaosGroup [19]. Material settings are in the Basic parameters rollout. When creating a scene with an airplane model, several materials with characteristic parameters were applied.

On the first stage we break the model into component parts of Element. The second stage is the creation of a material for imitation of plastic and applying it to the corresponding elements. Consider the basic parameters: The first rollout is Diffuse (color of the surface diffusion or diffuse color or just the main color). VRayMtl allows you to select a simple solid color for the Diffuse parameter or use a texture (map). In our case, we work, without texture, imitating painted metal. The second phase is a creation of plastic material for simulating and applying it to the respective n Elem there. The next rollout is the Reflection. As the name implies, this rollout contains reflective material properties. All surfaces in the world have reflective properties. If 100% black is chosen in this window, this deprives the surface of the material of reflections, white, on the contrary, makes it completely reflective. All intermediate gray values affect the power of reflections. The color slider in 3D Studio Max contains values from 0 to 255, and this means that to get a material with a reflectivity of 50%, you need to set the value to 128. In addition, you need to pay attention to the Fresnel IOR (index of refraction) setting. Fresnel IOR – Fresnel reflection coefficient (in other words, simply Fresnel reflections). This is a property that all objects in the world have. The essence of the effect lies in the fact that the larger the angle at which we look at the object, the weaker the reflections (Fig. 4.14).

The remaining materials for our plane are configured in the same way. We should also note the setting of the material of the upper deck with the texture of wood. In this case, the wooden flooring texture is used as a map for the Diffuse parameter. This is a previously prepared jpg file (Fig. 4.15).

The application of materials to the model occurs at the appropriate polygons. In the process of modeling, special polygons were foreseen for the corresponding materials. The further process is to sequentially select groups of polygons and apply the appropriate material to them. The next step is to set the parameters for lighting and final rendering of the object. Lighting is divided into two types: artificial (lamps) and natural (light from the sky and the sun). The most beautiful renderings are obtained by combining them. This lighting option is called mixed or combined. In order to make the image truly realistic, an external render program (visualizer) is used. VRay is the most popular program. This render remains the most popular, due to the fact that the most advanced methods are used in its calculations. It is based on the Monte-Carlo method. A method in which the process of lighting a 3d scene is modeled using a random value generator. Moreover, this program has additional technical advantages, giving it faster calculations, compared with comparables.

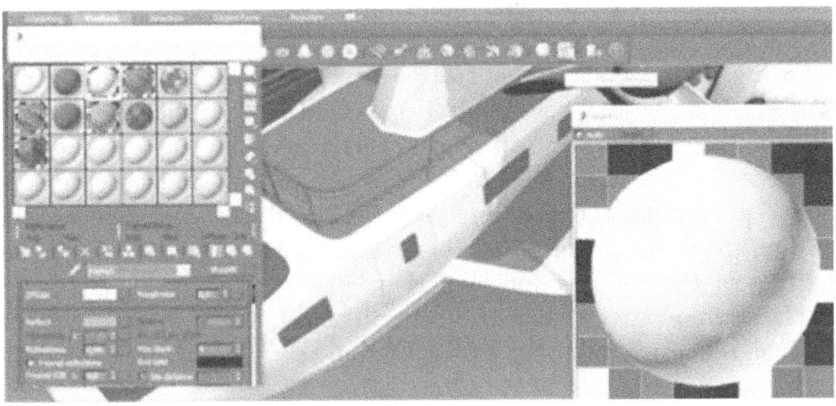

Fig. 4.14 An example of setting the material in the program dialog box.

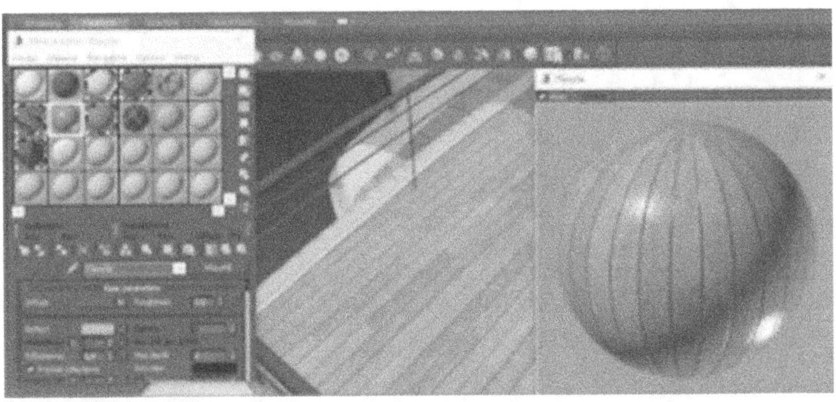

Fig. 4.15 An example of setting the material of the upper deck.

For our task, we will simulate studio lighting (artificial). To do this, a plane model is placed on the base without an obvious horizon line (the plane of the base smoothly passes into the plane behind). Thus, we will get a soft background, on which our model will be demonstrated (Fig. 4.16).

The presence of sources is the most important thing in our task. It should be noted that the number of light sources is individual in each new case; to simulate artificial lighting, at least two equivalent VRay-Light sources are necessary. Thus, we will surely get those very highlights on the surface of the objects, which is necessary for the naturalism of the final rendering. In the case of rendering of an amphibious plane, it is necessary to obtain a model image in the environment. The presence of a water surface is necessary for the final rendering (Fig. 4.17).

Getting to the final stage of our work, a number of points should be noted. For the final rendering, it is necessary to install cameras that will avoid some errors and in many respects shorten the working time. It is necessary to adjust the position of the camera and observe some rules of photography, for example, the rule of thirds. This rule assumes that the image should be divided into nine equal parts, and important elements of the composition should be placed along the intersections to create more interest than the central object. Fig. 4.17 a, b, c, show the final rendering of the scene with the new plane in the operating environment.

Fig. 4.16 An example of setting up materials and lighting of the scene with a model.

(a)

(b)

(c)

Fig. 4.17 Rendering of the plane model in the natural environment.

4.5 Conclusion

In conclusion, it can be noted that this paper presented the concept of a new multipurpose amphibious plane, which was based on the Consolidated PBY-5A Catalina prototype. The design process from the sketches to the final development of the final three-dimensional model and photo-realistic rendering is presented. This concept of amphibious aircraft is more spacious and can be indispensable as a multipurpose vehicle for the tourism industry and coastal waters patrols.

References

1. The military-historical portal /Website/Internet resource. Mode of access (date access 05/03/2017) www/URL: http://war20.ru.
2. Aviation portal /Website/Internet resource. Mode of access (date access 07/07/2017) www/URL: http://www.airforce.ru.
3. Information portal dedicated to the history of aviation /Website/Internet resource. Mode of access (date access 10/01/2017) www/URL: https://coollib.com
4. *Great Electronic Aviation Encyclopedia* /Website/Internet resource. Mode of access (date access 09/11/2017) www/URL: http://www.airwar.ru.
5. Official site TANTK them. Gm Beriev /Website/Internet resource. Mode of access (date access 01/06/2017) www/URL: http://www.beriev.com.
6. Petrov, G.F., *Amphibious and ekranoplanes of Russia in 1910–1999*, 248p, Rusavia, Moscow, 2000.
7. Howe, D., *Aircraft Conceptual Design Synthesis*, 474p, Professional Engineering Pub. Ltd, London, 2000.
8. Yeger, S.M., Matvienko, A.M., Shatalov, I.A., *Basics of aircraft: Textbook*, 720p, Mashinostroenie, M, 2003.
9. Jenkinson, L.R. and Marchman, J.F., *Aircraft design projects*, 371p, Butterworth-Heinemann, Oxford, 2003.
10. Vasin, S.A., Talaschuk, A.U. *et al.*, *Design and modeling of industrial products*, 692p, Mashinostroenie, Moscow, 2004.
11. Runge, V.F. and Manusevich, Y.P., *Ergonomics in environmental design*, 328p, Architecture-C, Moscow, 2005.
12. Raymer, D.P., *Living in the Future; The Education and Adventures of an Advanced Aircraft Designer*, 360p, Design Dimension Press, Los Angeles, 2009.
13. Gudmundsson, S., *General Aviation Aircraft Design*, pp. 521–545, Chapter 12. The Anatomy of the Fuselage, Elsevier, 2014.
14. Abbasov, I.B., *Computational modeling in industrial design*, 92p, DMK Press, Moscow, 2013.

15. Ahmadpour, N., Lindgaard, G., Robert, J.-M., Pownall, B., The thematic structure of passenger comfort experience and its relationship to the context features in the aircraft cabin. *Ergonomics*, 57, 6, 801–815, 2014. http://dx.doi.org/10.1080/00140139.2014.899632.
16. Abbasov, I.B., *A Fascinating Journey into the World of 3D Graphics with 3ds Max*, 239p, Amazon Digital Services LLC, 2017, ASIN: B076333ZQD [Electronic resource] https://www.amazon.com/Fascinating-Journey-into-World-Graphics-ebook/dp/B076333ZQD/ref=asap_bc?ie=UTF8.
17. Abbasov, I.B., Computational modeling of amphibious Be-200. *Proceedings of the Southern Federal University. Technical sciences*, 1, pp. 160–164, 2009.
18. Abbasov, I.B. and Orekhov, V.V., *Amphibious. Computational modeling*, 69p, LAP Lambert Academic Publishing, Saarbrucken, Germany, 2012, www.lap-publishing.com.
19. Mooney, T., *3ds Max speed modeling for 3D artists*, 422p, Packt Publishing, 2012.
20. Yablonsky, P.P., Cruise ships of the fatherland, 98 p. WIG of the world, Moscow, 1997.

5

Mathematical Model of Unmanned Aircraft with Elliptical Wing

Sergey A. Sinutin[1]*, Alexander A. Gorbunov[2] and Yekaterina B. Gorbunova[2]

[1]Southern Federal University, Engineering Technological Academy, Department of Embedded and Receiving Systems, Taganrog, Russia
[2]Southern Federal University, Engineering Technological Academy Technocenter, Taganrog, Russia

Abstract

The fifth chapter discusses the features of the formation of a mathematical model of an unmanned aircraft (UA) with elliptical wing. It is shown that such UAs occupy an intermediate niche between the UAs of the traditional aircraft scheme and the multirotor type UAs. Greater energy efficiency was achieved compared to multirotor type UAs and lower minimum flight speeds compared to UAs of traditional scheme. However, features of the formation of control and disturbing moments lead to the need for adaptive changes in the cost of the rudders depending on the flight speed and the value of the instantaneous angle of attack of the UA. The analytical method of research, as well as CFD-modeling are complex and do not provide sufficiently accurate results for the formation of a mathematical model. As a possible solution of the problem, the black box method is used. In order to collect the data needed to create a mathematical model of the control object using the black box method, a number of full-scale UA models with an elliptical wing were created; two versions of a special onboard recording device for recording and saving flight data were developed. Decryption of the information received is performed in a program created in the Matlab environment. The results of flight tests of scale models of UAs with elliptical wing are given. The mathematical model adaptation was performed according to the experimental data using the machine learning methodology, and the characteristic features of the control were identified.

**Corresponding author*: ssin@mail.ru

Iftikhar B. Abbasov (ed.) Computer Modeling in the Aerospace Industry, (179–210)

Keywords: Unmanned aircraft, control system, elliptical wing, model-oriented design, simulink, roll angle, pitch angle, mathematical model

5.1 Introduction

In recent years, a qualitative leap in the development of unmanned aircraft throughout the world can be observed. The relevance of research in this area is beyond doubt, since unmanned aircrafts (UAs) have proven to be cost-effective and effective means of solving problems of aerial photography, mapping, environmental monitoring, and several others. The possibility of using UAs in emergency areas without risking the life of the operator is also attractive. In the modern engineering community, a trend has emerged to develop and study new types of aerodynamic schemes for aircraft, a reflection of which is a number of published works [1, 2]. New concepts of aerodynamic configurations are needed to optimize the properties of the device to the area of its use, while the availability of the developers of modern hardware and software design, debugging and verification of onboard radio-electronic systems allows the exploration and application of schemes not previously used in connection with the complexity of management.

In this sense, the choice of the elliptical shape of the UA wing depends on the specifics of its aerodynamic properties, which could not be better suited for solving the problem of aerial photography of the area remote from the point of basing. These properties consist in the possibility of implementing two flight modes: fast and energy efficient, necessary to get to a destination, and steady slow to make high-quality photography of the point of interest. At the same time, the specificity of the aerodynamics of an elliptical wing determines a number of control features of such an apparatus that need to be studied.

5.2 Research Objective

The objective of this research work is to create a flight control system for UA with an elliptical wing with the ability to maintain a given speed, flight direction and attitude. The purpose of the first stage of research and development is to study the characteristic features of the behavior of UA with an elliptical wing in the air and its reactions to control actions. Previous patent researches showed the presence of similar developments, but no complete analogs were found.

The scientific novelty of the research topic consists not only in obtaining the laws of controlling the apparatus of a non-standard aerodynamic scheme, but also in developing a methodology for solving such problems on the basis of the model-oriented approach.

5.3 Research Technique

The analytical method of research, as well as CFD modeling, do not provide sufficiently accurate results for the formation of a mathematical model and, therefore, the control laws [3]. The solution in this case is the application of the black box method, when instead of studying the properties and interrelationships of the system's components, its reaction to changing conditions as a whole is studied. In this case, changing conditions should be understood as control commands from the ground control of a human operator, and the reaction is the change in the position of the unmanned aircraft in space [4].

Literature analysis showed the widespread use of the black box method for solving the problem of studying the behavior of non-stationary objects, for which a special on-board recording device (BRD) was assembled on the basis of an Arduino Uno debug board with an AtMega328 microcontroller. The evaluation of the output reactions of the UA was carried out using data from the accelerometer ADXL 325. The data from the input and output of the system is recorded on the SD card [5, 6].

After writing and debugging code based on the standard Arduino Uno libraries, the BRD was reassembled around the Arduino Mini Pro, which is more suitable for placing onboard a UA due to its small size and weight. Fig. 5.1 shows an example of a timing diagram of signals recorded by the BRD during flight tests. At this stage, the need to introduce an additional sensitive element into the system was revealed – a gyroscope, without which it is extremely difficult to correctly determine the spatial position of the UA.

5.4 Hardware Implementation

For the further development of the project, a second version of the BRD was developed on the STM32F 4 platform. The need to change the hardware platform was caused by the inability to simultaneously poll sensitive elements by interrupting and recording the received data on a solid-state drive. In addition, the expansion of the functionality of the BRD to the control system involves the installation of a real-time operating system (RTOS), while the Arduino Mini Pro does not provide sufficient

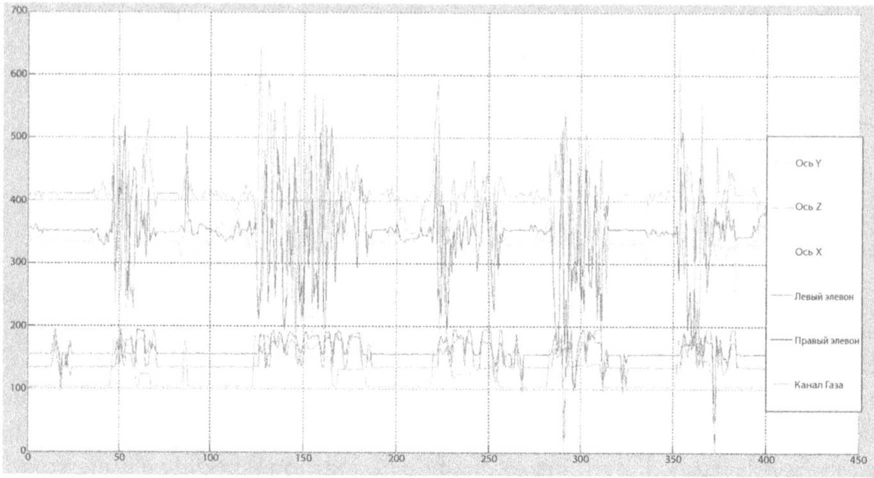

Fig. 5.1 Graphic representation of raw flight information.

performance for this [5, 6]. The following elements were used as sensitive elements: a 3-axis accelerometer LIS302DL and a 3-axis angular velocity sensor (DLS) L3GD4200, connected via an I2C interface (see Fig. 5.2).

Sensor readings and data on control signals from the ground operator's console are stored in text format on the SD card. At this stage, all flight tests were conducted in manual control mode.

Then, MBee-868-2 868 MHz radio module was installed on the BRD for use in wireless data transmission and control systems, industrial

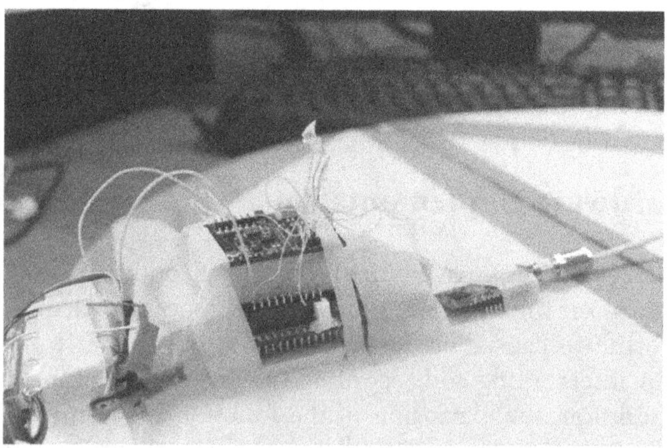

Fig. 5.2 General view of the recording device based on STM32F4.

telemetry and security systems. After writing the appropriate PC host program, this made it possible to obtain onboard sensor readings during flight tests in real time.

The MBee-868-2 modules are based on the System Instruments-on-Chip family of CC430 chips from Texas Instruments and support the 6LoWPAN and SimpliciTI protocols. The used frequency range, and high output power provides stable communication at a distance of tens of kilometers in the line of sight, allowing you to deploy systems without installing additional repeaters.

5.5 The Program Research Part

With he increasing complexity of the project it became necessary to attract more advanced technologies for writing and debugging the program part. At this stage, the Model-Oriented Design Toolkit (MOS) was mastered and applied [7].

This approach is characterized by a high degree of visualization of signal processing and the design of complex control systems. At the heart of the development process is a systemic behavioral model consisting of block diagrams, text programs and other graphical elements, which is essentially an executable specification used for designing, analyzing, simulating, automatically generating and verifying the project code throughout the development process. Thus, to create the program part of the second version of the BRD, the Matlab/Simulink package was used, which implemented the MOS paradigm (see Fig. 5.3) [8].

Using the system model as a specification provided the following benefits:

- testing, refining and retesting of the project during the entire development process;
- the possibility of simulating and validating the code of the BRD program before the microcontroller firmware;
- the possibility of representing the entire system as a hierarchical structure;
- automatic code generation from the system model, which reduces the number of errors and significantly speeds up the development and debugging process;
- the ability to quickly edit and adapt models;
- creation of embedded software (software) and applications for a personal computer used as a ground control station in a single development environment (see Fig. 5.4 and 5.5).

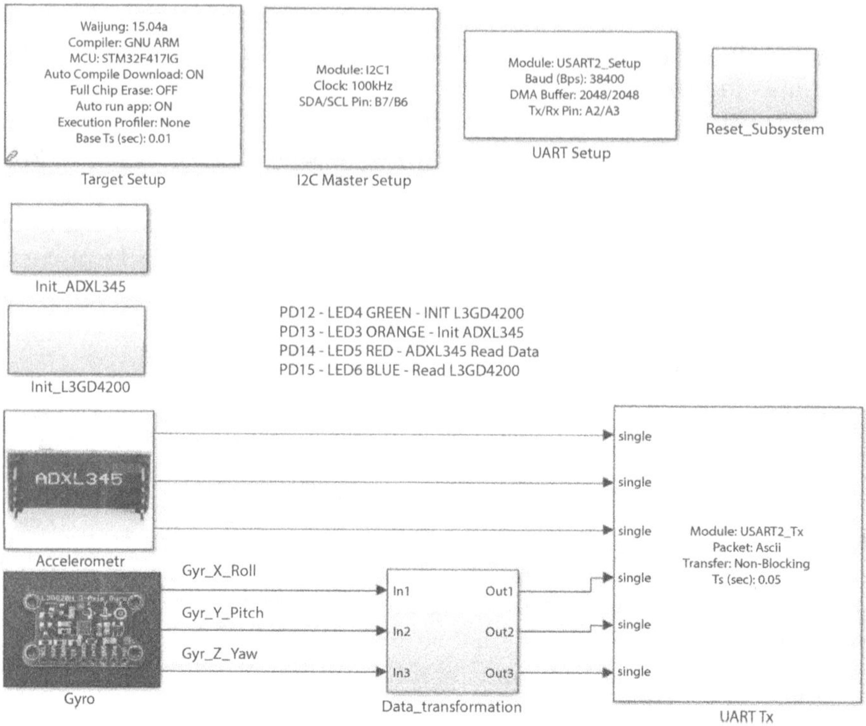

Fig. 5.3 General view of the model of functioning of the onboard recording device of the UA, built on the principles of MOS.

It should be noted that when creating the BRD system model, not only libraries provided by MathWorks were used, but developed by third-party companies (Aimagin — RapidSTM and Waijung Blockset libraries) [9], including developers of electronic components (STM - STM32- Mat Target, PIC - Microchip-Simulink Blockset). The use of MOS allowed a reduction in the final development time and debugging of the joint operation of the onboard and ground parts of the system [10].

5.6 Studies of the Behavior of an Unmanned Aircraft with an Elliptical Wing

The elliptical wing shape of the UA was chosen due to its specific aerodynamic properties [1, 11]. Its advantage over the classical aircraft scheme is the possibility of stable flight at ultra-low speeds. However, it is much

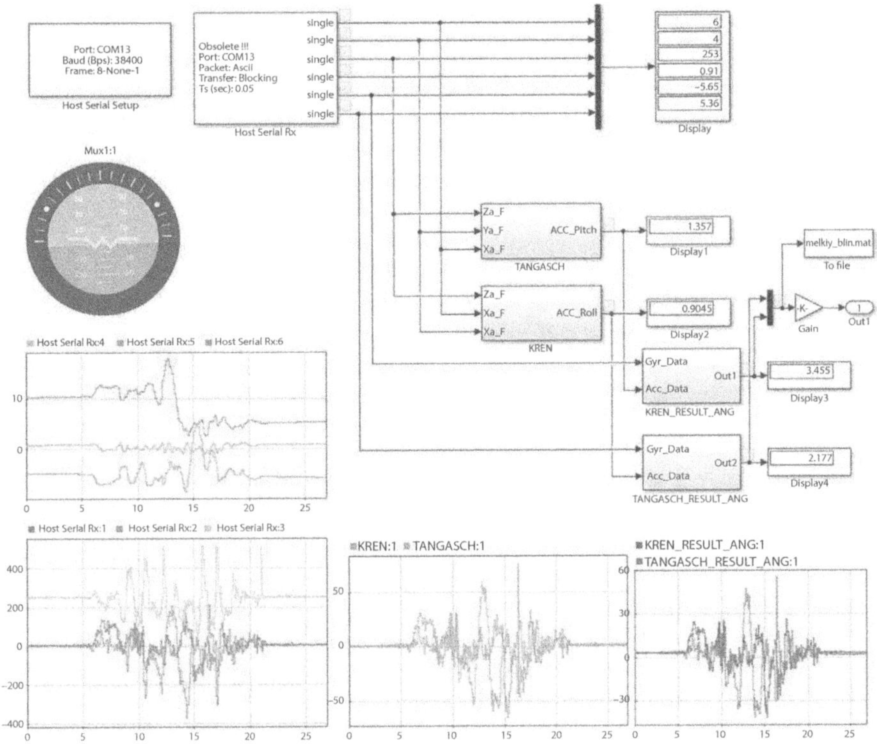

Fig. 5.4 General view of the flight data analysis host.

more energy efficient than a multirotor. On the other hand, the specificity of the aerodynamics of an elliptical wing requires research to form correct control laws for such UAs [4].

When purging a virtual model of the selected type of UA in the XFLR5 wind tunnel emulator program, it was found that at a speed of 4 m/s and angles of attack up to 10^0 , the wing flute is 12^0 (see Fig. 5.6), and the total aerodynamic force moment, applied in focus, is negative, which indicates the likely appearance of a dive moment during takeoff. The results of the change in the flux at a constant speed and different angles of attack of the wing are shown in Fig. 5.6–5.8.

During the flight tests, an increased sensitivity of this aerodynamic scheme to the pitch channel control, relative to the roll channel, was detected, as well as a pronounced dive moment, especially noticeable during takeoff [4].

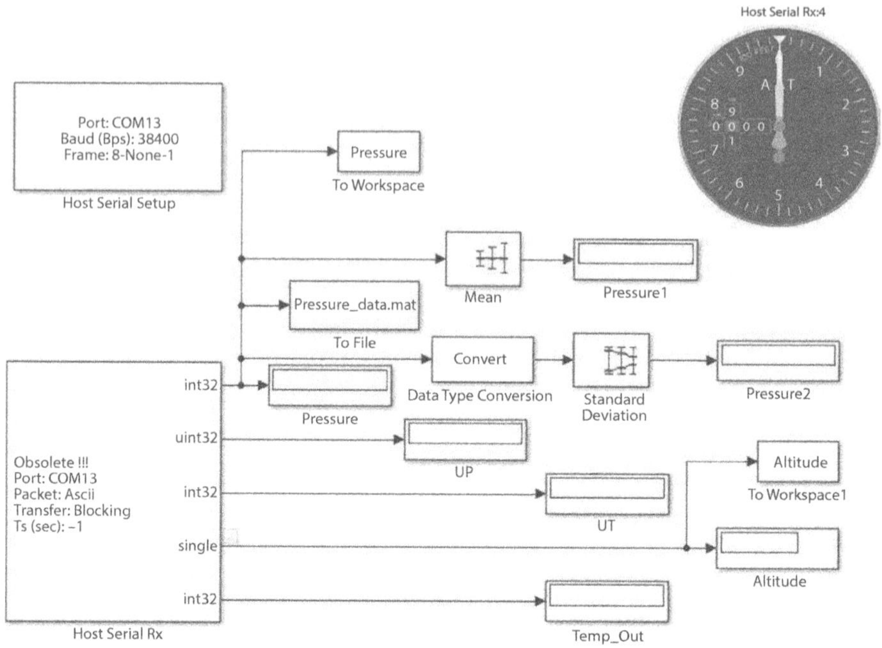

Fig. 5.5 The subsystem for determining the altitude of the flight of the UA according to the pressure sensor.

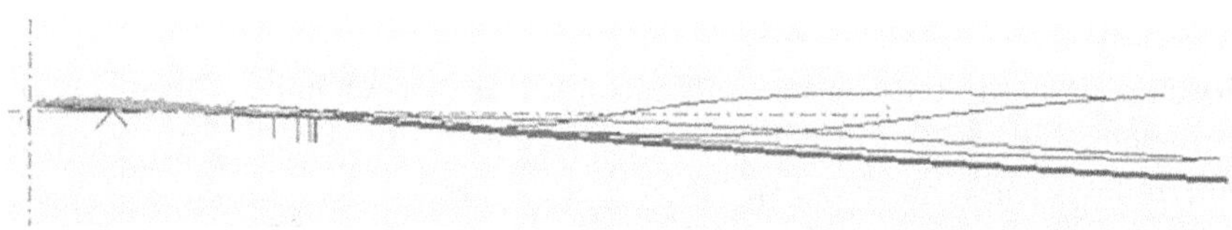

Fig. 5.6 Wing flow slant at a speed of 4 m/s and an angle of 10°.

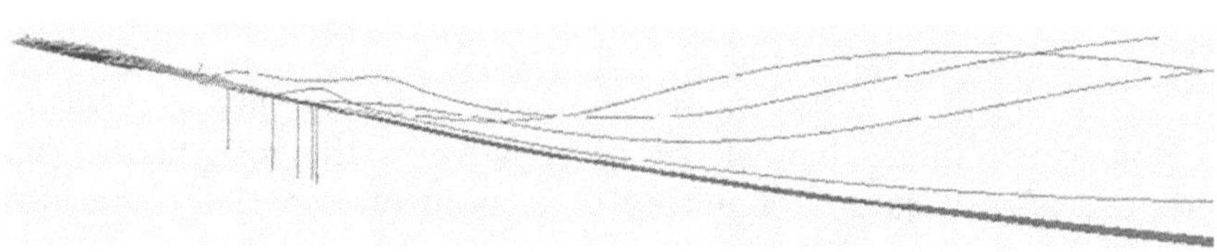

Fig. 5.7 Wing flow slant at a speed of 4 m/s and an angle of 10°.

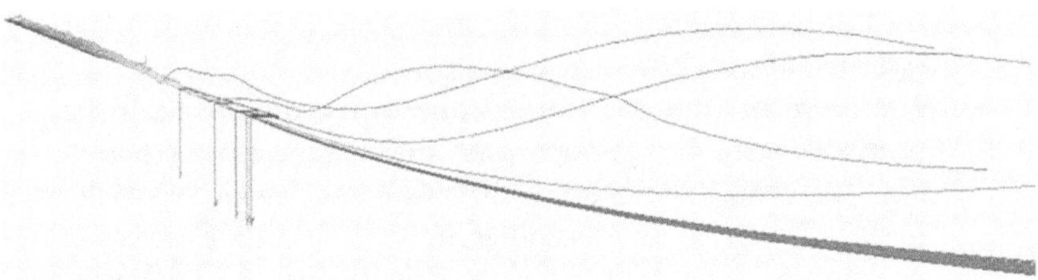

Fig. 5.8 Wing flow slant at a speed of 4 m/s and an angle of 20°.

5.7 Experimental Studies of the UA Behavior

1. To optimize energy consumption in the mode of small horizontal flight speeds with large angles of attack, the power plant was chosen taking into account the use of a large-diameter propeller. For one of the studied UAs, the diameter of the propeller was 65% of the wing span (see Fig. 5.9) [5, 6]. This unit was subject to "twisting", especially at the time of takeoff. To compensate for the torque, an adaptive control over the roll channel relative to the horizontal speed of the UA was used (see Fig. 5.9).
2. The UA with an elliptical wing has two flight modes: FFF (Forward Fast Flight) and FSF (Forward Slow Flight). For FFF, the power requirement is 25%. For the FSF, this figure

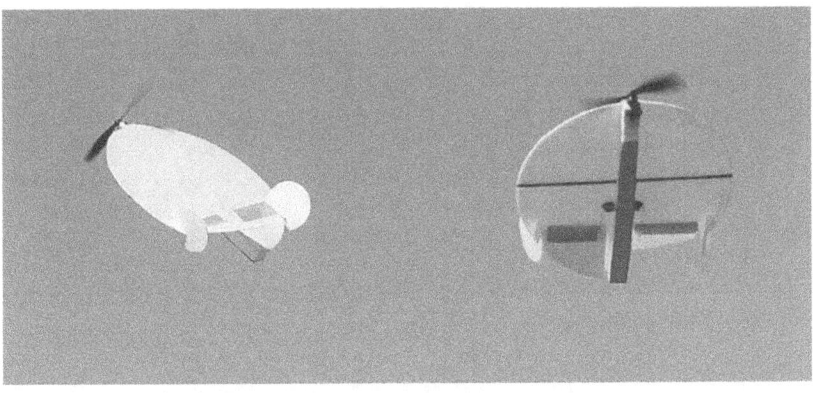

Fig. 5.9 Compensation of the torque generated by the propeller of a single-engine UA with an elliptical wing, by deflecting the air rudders.

is (35%–65%), which is much lower than that of multirotor systems (at least 100%).

3. The moments that are formed due to the difference of the center of mass and pressure become insufficient for parrying the possible effects on pitch. As a result, the device in the sense of pitch control becomes neutral, and then statically unstable. Such a situation is especially dangerous in the presence of unsteady disturbing moments.
4. The controllability of the roll channel is neutral, but with decreasing speed the control torque decreases. The solution to this problem is to increase the area of aerodynamic control surfaces, which necessitates adaptive control due to the redundancy of the moment in flight mode FFF.
5. Forced airflow control surfaces showed the effectiveness of this solution. In Fig. 5.10 the interposition of aerodynamic control surfaces and power units are shown.

It should be borne in mind that the LA model is not ideally smooth; therefore, the aerodynamic control surfaces operate in a noise air flow. Additional high-frequency effect on the rudders has a disturbed flow from the screws of the propulsion system [4].

Fig. 5.10 Location of UA elevons with elliptical wing.

5.8 Processing and Analysis of Data Obtained during Flight Tests

Flight tests, accompanied by the fixation of telemetry, are necessary to identify features of the control apparatus and the subsequent identification of the mathematical model of its behavior. Flight information processing was carried out on a PC using specially created software.

To determine the pitch and roll angles based on accelerometer readings, the following ratios were used [8]:

$$\beta = arctan\left(\frac{A_y}{\sqrt{A_x^2 + A_z^2}}\right) \tag{5.1}$$

$$\psi = arctan\left(\frac{A_z}{\sqrt{A_x^2 + A_y^2}}\right) \tag{5.2}$$

The constant component was compensated. Fig. 5.11 shows example of the representation of the processed flight information.

In the first versions of the BRD, the flight information was recorded only on the built-in SD card and processed in the laboratory. In the last modification, the option of data transmission over the air channel to the ground station was added to the BRD. In addition to the rapid detection of data recording problems, this made flight tests more informative due to the possibility of observing telemetry in real time, but required processing of host programs [4].

To determine the spatial position of the UA, the accelerometer readings (see Fig. 5.12) were used in conjunction with the angular velocity data. For this purpose, the system uses a three-axis angular velocity sensor (CRS). Integration of the angular velocity sensor (TLS) over time gives the corresponding angles, but the result always has a certain trend due to the accumulation of integration error (Fig. 5.13). Accounting for integration errors allows the normalization of the data on angular velocities (Fig. 5.14).

To filter the data, a first-order digital inertial link with coefficient $\alpha = 0.02$ was used. The use of accelerometer and TLS data in combination with the use of a complementary filter with a coefficient $K = 0.05$ allowed us to determine the required angular position of the aircraft in space with the

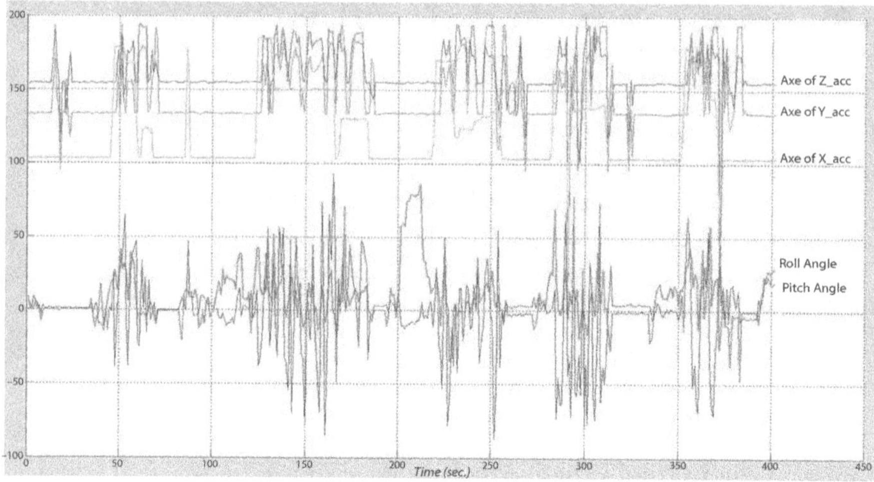

Fig. 5. 11 Input effects and orientation angles of the UA in space during flight.

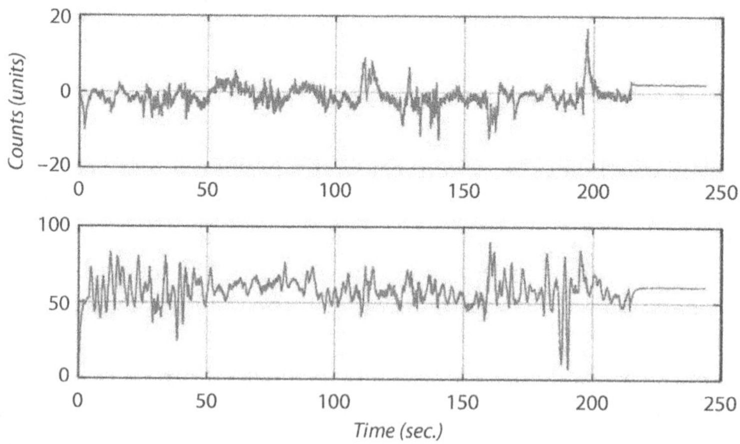

Fig. 5.12 Accelerometer data after filtering.

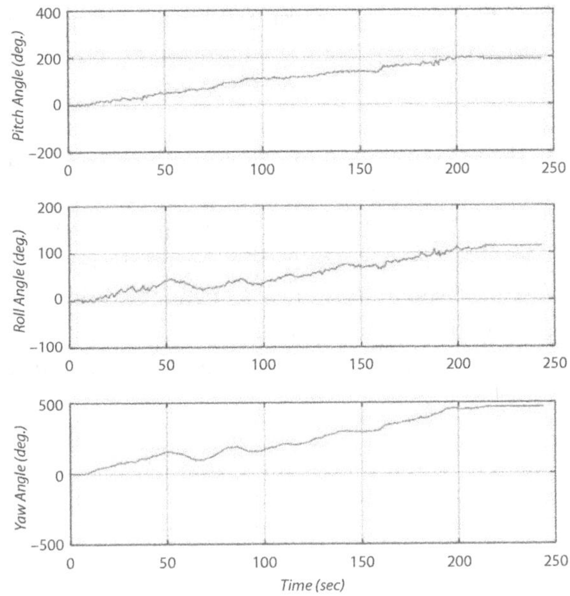

Fig. 5.13 The result of integrating CRS data.

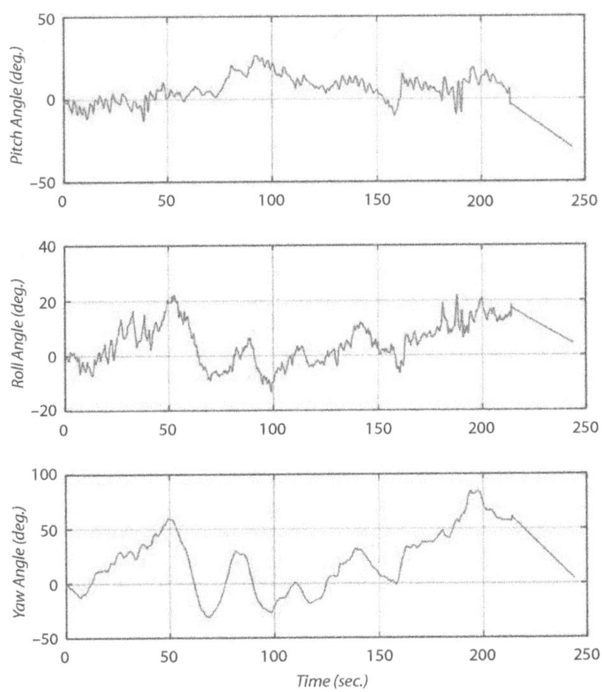

Fig. 5.14 The data on the angular velocity with the trend

required accuracy, therefore, the response of the system under investigation to control actions (see Fig. 5.15.). Validation of the data interpretation was carried out using time synchronization of the obtained graphs with video recording of UA flight.

For the formation of a mathematical model, it is necessary to supplement these data with information on the height and linear velocity of flight. As mentioned above, the new version of the flight data processing program was created in the basis of model-based design, and provides for the collection and analysis of flight data in quasi-real time.

The "to file" block allows you to save flight data in a file with the .mat extension for further analysis, and building graphs of the change in the

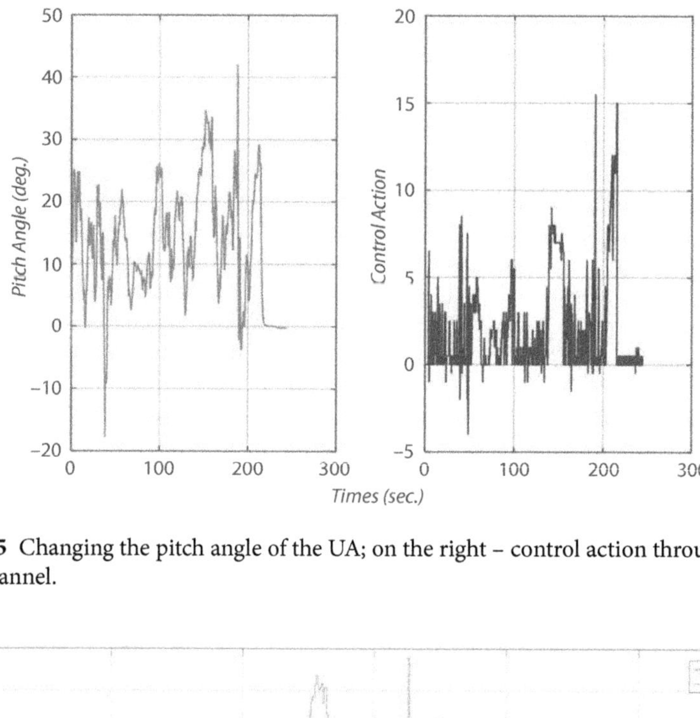

Fig. 5.15 Changing the pitch angle of the UA; on the right – control action through the pitch channel.

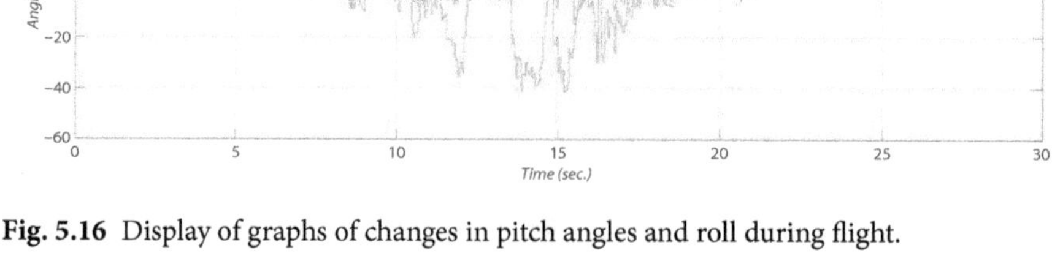

Fig. 5.16 Display of graphs of changes in pitch angles and roll during flight.

angular position of the UA during the flight can be done with a few clicks (an example is in Fig. 5.16).

5.9 Formation of a Mathematical Model of UA with Elliptical Wing

Within this R&D, two classical control problems are solved: the stabilization problem and the navigation problem [12]. A mathematical model of the control object is necessary to predict its response to control actions and environmental changes. To identify a mathematical model, two approaches are considered: analytical, based on Newton's second law, and the black box method, which makes it possible to approximate the transfer function of the system using experimental data.

5.10 Mathematical Model of UA in Analytical Form

The basis of the study of the motion of objects in the macrocosm is Newton's second law. Speaking of body movement, it implies its movement in space and in time relative to the reference system. Taking the mechanical model of a UA as an absolutely solid body with constant mass, the motion will be described by the following equations [13]:

$$\begin{cases} m\dfrac{d\vec{V}}{dt} = \vec{F}; \\ \dfrac{d\vec{K}}{dt} = \vec{M}; \end{cases} \tag{5.3}$$

where $\vec{V}$ – vector of speed of movement; $\vec{F}$ – total vector of external forces acting on the UA; m – UA mass; $\vec{M}$ – total vector of external force moments. The system (5.3), presented in vector form, determines the change in the kinematic parameters of the UAV motion under the action of specified external forces. Studies of the flight conditions of the UA, its stability and controllability are carried out with the help of the system of differential equations (5.4), which can be considered as its mathematical model.

$$\begin{cases} m\left(\dfrac{dV_x}{dt}+\omega_x V_z-\omega_z V_y\right)=F_x \\ m\left(\dfrac{dV_z}{dt}+\omega_x V_x-\omega_x V_z\right)=F_y \\ m\left(\dfrac{dV_z}{dt}+\omega_x V_y-\omega_y V_x\right)=F_z \end{cases}$$

$$\begin{cases} I_x\dfrac{d\omega_x}{dt}-I_{xy}\left(\dfrac{d\omega_y}{dt}-\omega_x\omega_z\right)+(I_z-I_y)\omega_y\omega_z=M_x \\ I_y\dfrac{d\omega_y}{dt}-I_{xy}\left(\dfrac{d\omega_x}{dt}-\omega_y\omega_z\right)+(I_x-I_z)\omega_x\omega_z=M_y \\ I_z\dfrac{d\omega_z}{dt}+(I_y-I_y)\omega_x\omega_y+I_{xy}\left(\omega_y^2-\omega_x^2\right)=M_z \end{cases} \tag{5.4}$$

$$\begin{cases} \omega_x=\dfrac{d\gamma}{dt}+\dfrac{d\varphi}{dt}sin\vartheta \\ \omega_y=\dfrac{d\varphi}{dt}cos\vartheta cos\gamma+\dfrac{d\vartheta}{dt}sin\gamma \\ \omega_z=\dfrac{d\vartheta}{dt}cos\gamma-\dfrac{d\varphi}{dt}cos\vartheta sin\gamma \end{cases}$$

$$\begin{cases} m\left(\dfrac{dV_x}{dt}+\omega_x V_z-\omega_z V_y\right)+mg\sin\vartheta-C_x-F_x^T=0 \\ m\left(\dfrac{dV_y}{dt}+\omega_x V_x-\omega_x V_z\right)-mg\cos\vartheta\cos\gamma-C_y-F_y^T=0 \\ m\left(\dfrac{dV_z}{dt}+\omega_x V_y-\omega_y V_x\right)-mg\cos\vartheta\, sin\,\gamma-C_z-F_z^T=0 \end{cases} \tag{5.5}$$

Let us rewrite the first 3 equations of the system (5.4), by writing and transferring the forces acting on the aircraft to the left side:

Here is a vector $\vec{C}$ is a set of aerodynamic forces acting on the device, $\overrightarrow{F^T}$ is engine generated thrust. Such a mathematical model is common for aircraft, in which the lift force is created by a wing. Wing geometrydetermines vector $\vec{C}$, the values of which for devices with non-standard geometry should be estimated by blowing in the wind tunnel.

5.11 Obtaining a Mathematical Model Using the "Black Box" Method

To implement the black box method, the Matlab extension of the System Identification Toolbox (Fig. 5.17) was used [13]. With it, an attempt was made to obtain the transfer function of the system through the pitch channel using experimental data obtained by the switchgear.

Fig. 5.18 shows a fragment of the timing diagram of the pitch angle change, and Fig. 5.19 shows the corresponding control action. The result of the System Identification Toolbox program is shown in Fig. 5.20. This extension allows us not only to make an assessment of the transfer function, but also to assess the degree of its coincidence with the prototype. The prototype is a real change in the pitch angle, which was registered by the onboard equipment.

The analytical form of the transfer characteristics in the pitch channel was determined for several cases.

The transfer function of the channel pitch UA with an elliptical wing with a coincidence of 91.95% is:

$$f_{pitch} = \frac{0.03395z^{-1} - 0.01893z^{-2}}{1 - 2.819z^{-1} + 3.506z^{-2} - 2.233z^{-3} + 0.5494z^{-4}} \tag{5.6}$$

The transfer function of the UA pitch channel with an elliptical wing with a coincidence of 90.47%:

$$f_{pitch} = \frac{0.003168z^{-1}}{1 - 2.886z^{-1} + 2.846z^{-2} - 0.9601z^{-3}} \tag{5.7}$$

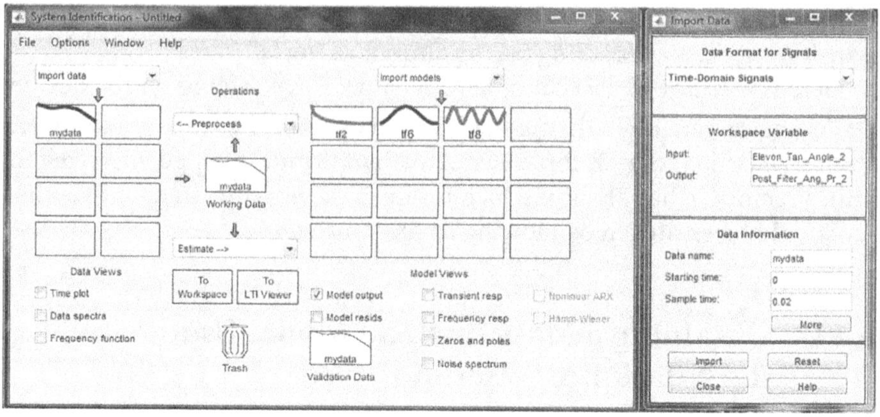

Fig. 5.17 View system independence toolbox subroutine.

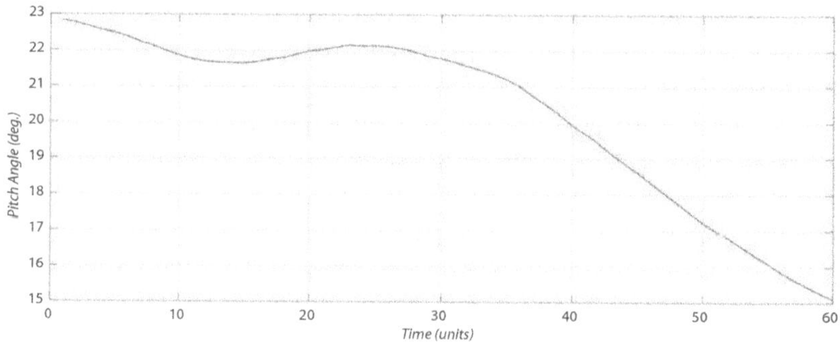

Fig. 5.18 Diagram of changes in the pitch angle of the UA.

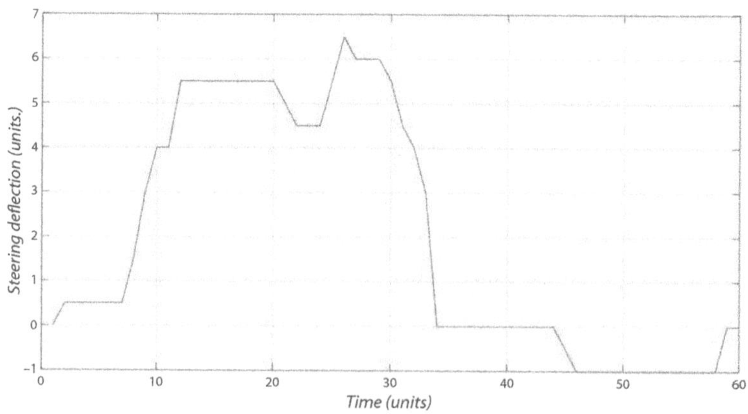

Fig. 5.19 Diagram of changes in the impact on the investigated UA.

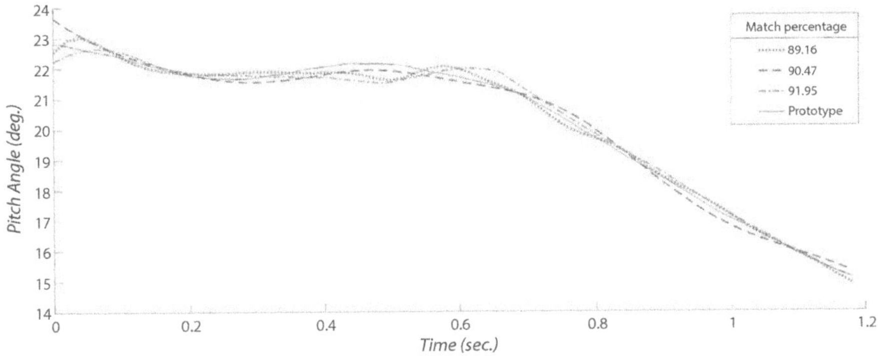

Fig. 5.20 The result of determining the transfer function of the pitch channel.

The transfer function of the UA pitch channel with an elliptical wing with a coincidence of 89.16%:

$$f_{pitch} = \frac{3.363s^2 + 32.2s + 146.2}{s^3 + 8.531s^2 + 57.37 + 27.35} \tag{5.8}$$

5.12 Mathematical Model Based on Linear Regression

The UA under study is made according to the tailless scheme, for which the critical problem is control over the pitch channel [5, 6], so the study was focused in this area.

An unmanned aircraft is a complex non-linear system operating in a weakly deterministic medium, and therefore its analytical description presents a known complexity. Fortunately, the development of modern computing and numerical analysis contributed to the development of a whole class of methods called "machine learning methods" [15], which involve adapting the model by "learning" on empirical data. At the first stage, a second-order linear regression model was tested:

$$\text{Pitch}_{\text{angle}} = A \times X, \tag{5.9}$$

where A– model coefficient vector, X– input parameter vector:

$$A=\begin{bmatrix} a_0 \\ a_1 \\ \dots \\ a_{20} \end{bmatrix}; X=\begin{bmatrix} 1 \\ \text{Left}_{\text{el}} \\ \text{Right}_{\text{el}} \\ \text{Thr} \\ V \\ H \\ \text{Left}_{\text{el}}^2 \\ \text{Right}_{\text{el}}^2 \\ \text{Thr}^2 \\ V^2 \\ H^2 \\ \text{Left}_{\text{el}} \cdot \text{Right}_{\text{el}} \\ \text{Left}_{\text{el}} \cdot \text{Thr} \\ \text{Left}_{\text{el}} \cdot V \\ \text{Left}_{\text{el}} \cdot H \\ \text{Right}_{\text{el}} \cdot \text{Thr} \\ \text{Right}_{\text{el}} \cdot V \\ \text{Left}_{\text{el}} \cdot H \\ \text{Right}_{\text{el}} \cdot \text{Thr} \\ \text{Right}_{\text{el}} \cdot V \\ \text{Right}_{\text{el}} \cdot H \\ \text{Thr} \cdot V \\ \text{Thr} \cdot H \\ V \cdot H \end{bmatrix}, \tag{5.10}$$

where Left_{el} – the angle of the left elevon; Right_{el} – the angle of the right elevon; Thr – gas; V – ground speed; H – flight altitude.

The training sample was formed from data collected during flight tests; training of the model was carried out by the gradient descent method; to accelerate the gradient descent, all parameters (including polynomial) were normalized in accordance with formula (5.11):

$$x_{\text{H}} = \frac{x - \bar{x}}{\max(x)} \tag{5.11}$$

In addition to the loss function, the coefficient of determination (4) was used to control the model's quality:

$$R^2 = 1 - \frac{\sum_{i=1}^{l} \left(h(x_i) - y_i\right)^2}{\sum_{i=1}^{l} \left(y_i - \bar{y}\right)^2} \tag{5.12}$$

The better the model describes the data, the closer the value R^2 to 1. Fig. 5.21 compares the pitch angle log and the model forecast.

Table 5.1 shows the values of the coefficients of the model.

The dependence in Fig. 5.21 demonstrates the correlation with the real log data, but does not take into account all their features. Thus, the

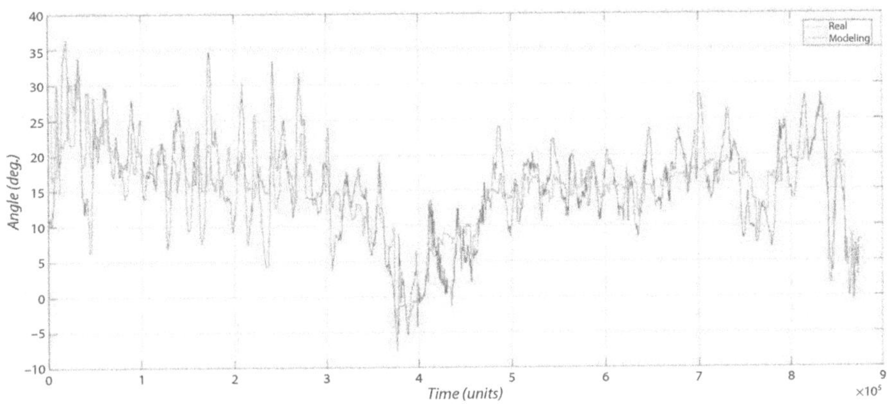

Fig. 5.21 Pitch angle prediction using a linear regression model ($R^2 = 0.6$).

Table 5.1 The values of the coefficients of a linear regression model of the second order after denormalization.

a_0	46.69	a_7	2.55	a_{14}	0.0338
a_1	-14.24	a_8	0.35	a_{15}	-0.746
a_2	-23.20	a_9	-0.026	a_{16}	-0.3074
a_3	12.6	a_{10}	-0.0037	a_{17}	-0.157
a_4	5.82	a_{11}	0.2285	a_{18}	-0.025
a_5	0.79	a_{12}	-1.1156	a_{19}	0.0143
a_6	1.5	a_{13}	-0.502	a_{20}	0.0216

first approximation of the mathematical model for the pitch channel was obtained.

5.13 Mathematical Model Based on Multilayer Perceptron

Models based on artificial neural networks (NN), have proven themselves in describing data with a complex non-linear structure. The choice of architecture and metaparameters of the neural network, as a rule, has a heuristic character. Neural networks with five neurons in the first layer were considered that corresponded to the data of the right elevon, left elevon, the position of the gas knob, the velocity relative to the ground and the height; one hidden layer and one neuron in the output layer. Table 5.2 presents data on the standard deviation of the model prediction from real data and the coefficient of determination (4) for NN, which has 10 neurons in the hidden layer with different learning methods.

The Levenberg-Marquard method [15] in this case allowed achieving the best quality of the data approximation. At the same time, the accuracy of the forecast was not much higher than that achieved using a linear regression model. Since the presence of a prototype UA together with the onboard recorder made it possible to collect any required amount of data, it was possible to follow the path of increasing the number of neurons in the hidden layer without the danger of retraining the network.

Table 5.3 presents data for NN, trained by the Levenberg-Markvard method, with 10, 25 and 50 neurons in the hidden layer. The value of the coefficient of determination achieved for the NN, which has 50 hidden neurons, indicates a very high quality of the approximation of data by a neural network.

Fig. 5.22 shows the time diagram to which the real log data is compared with the output from the NN.

Table 5.2 Indicators of quality of data approximation with different methods of teaching NN.

	Levenberg-Marquardt	**Bayes method**	**Conjugate gradient method**
SKO	2.847	2.99	3.16
R^2	0.67	0.598	0.5626

Table 5.3 Indicators of quality approximation of data at different sizes of the hidden layer.

The number of hidden neurons	**10**	**25**	**50**
SKO	2.847	1.9247	1.11
R^2	0.67	0.858	0.9549

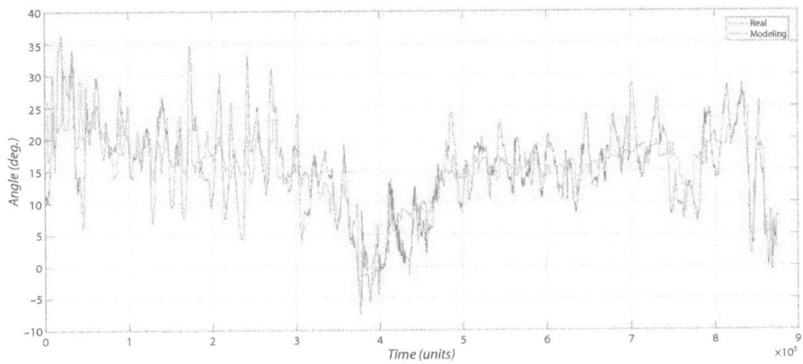

Fig. 5.22 Pitch angle forecast using the neural network model ($R^2 = 0.95$).

In Fig. 5.22, it is clear that the model indeed quite accurately approximates real data. Emissions at some points, the nature of which is currently not clear, are surprising. Probably, they are associated with errors in writing input parameters. Since the emissions are pulsed, at this stage it is possible to eliminate them with a digital filter.

5.14 PID Controller Setup

For the convenience of tuning the PID controller, the obtained mathematical model of the UAV was exported via the pitch channel to Simulink, where the system model was assembled (Fig. 5.23).

At the entrances responsible for the gas channel, the speeds and altitude of the flight were given constant values, the initial initialization of the rudders was fed to the mechanization control channel [14]. The output of the system was compared with the desired angular position,

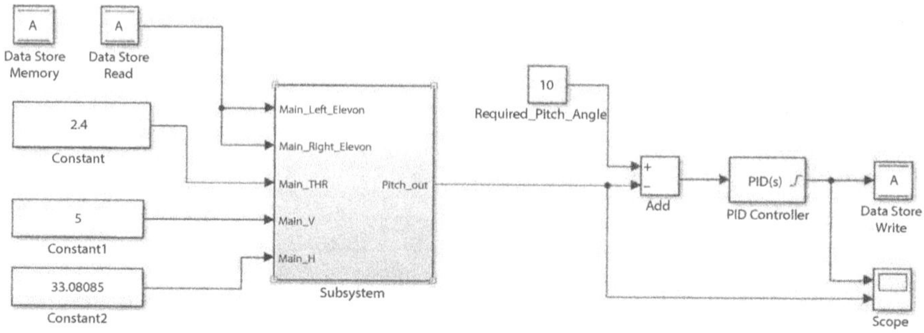

Fig. 5.23 Simulation model of the pitch channel to determine the coefficients of the PID controller.

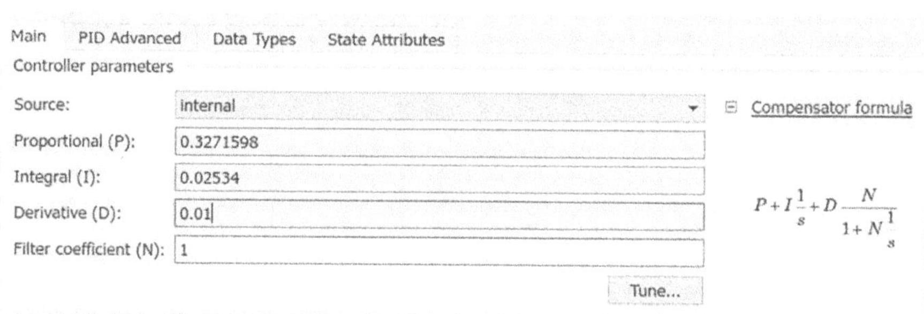

Fig. 5.24 PID coefficients.

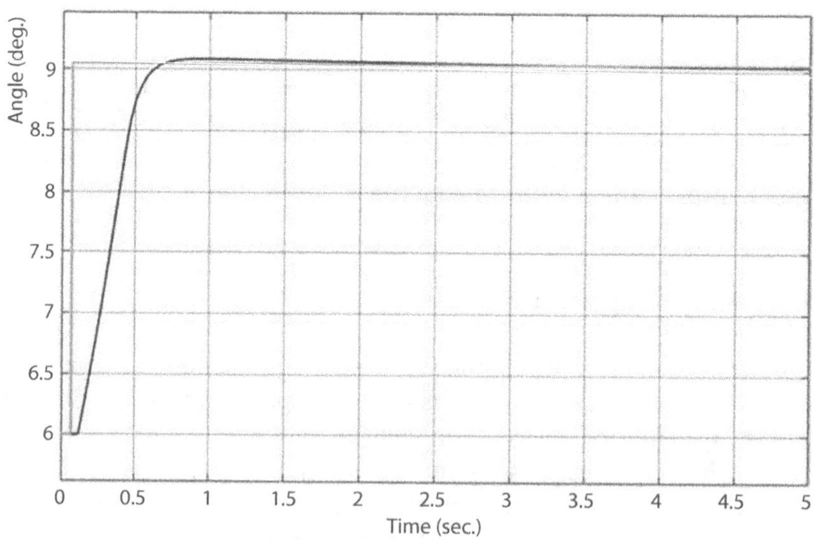

Fig. 5.25 Transition function of the system with an abrupt change in pitch angle.

the result was fed to the PID controller unit, the output values of which were limited to approximate real values. For convenience of debugging, values from the feedback line were fed to the system through the Data Store block.

The coefficients of the PID controller, which provide the best stability without overshooting with the minimum reaction time (Fig. 5.24), are presented in Fig. 5.25.

5.15 Flight Emulation for Primary Quality Control of the Regulator

The initial assessment of the quality of the controller was carried out using a flight simulator. This made it possible to avoid damage to the existing prototype UA with an elliptical wing. The following requirements were made for flight simulator programs:

- the ability to create your own computer model of the aircraft;
- the ability to interact with Matlab/Simulink.

Two simulators met these requirements: FlightGear and X-Plane. Simulink has a unit that provides communication with FlightGear. The interaction of the X-Plane flight simulator with Matlab is possible after installing additional free libraries distributed by the user community. For interconnecting flight simulators with the simulation environment, the UDP protocol was used.

The choice was made in favor of the X-Plane program, since the built-in Plane Maker tool has a graphical interface and allows you to create models of complex shape, which is extremely important, and in the case of a circular wing, it is not feasible with the parametric description adopted in JSBSim used in the development of FlightGear simulator models.

X-Plane differs from other simulators using the blade element theory developed to evaluate the behavior of the propeller. Each element of the aircraft is divided into small sub-elements (see Fig. 5.26–5.29), for which the existing aerodynamic forces and moments are calculated, then they are all added up to obtain a general result. This approach applies to all elements that create lift.

The advantage of this method compared to models using pre-calculated data arrays is that the X-Plane simulator itself simulates the behavior of the aircraft, based on data about the geometry, mass, location and

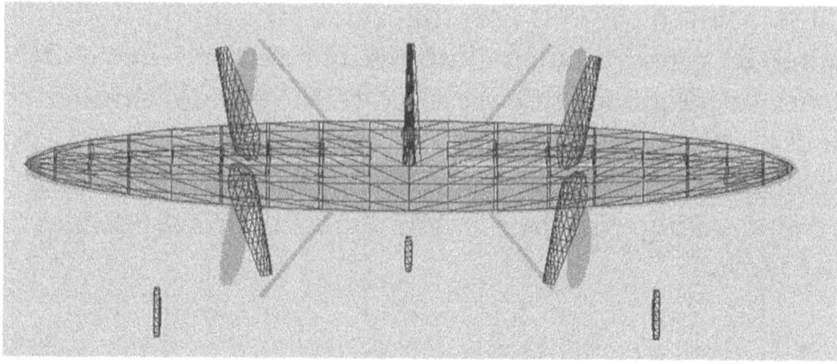

Fig. 5.26 UA model with elliptical wing, designed for the X-Plain flight simulator: front view.

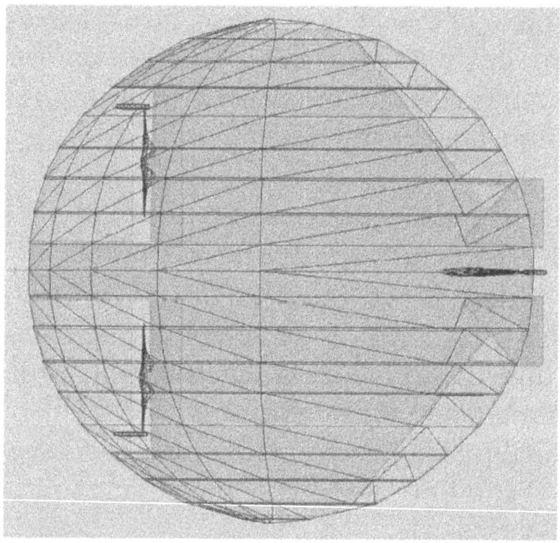

Fig. 5.27 Model of UA with elliptical wing, created for the X-Plain flight simulator: top view.

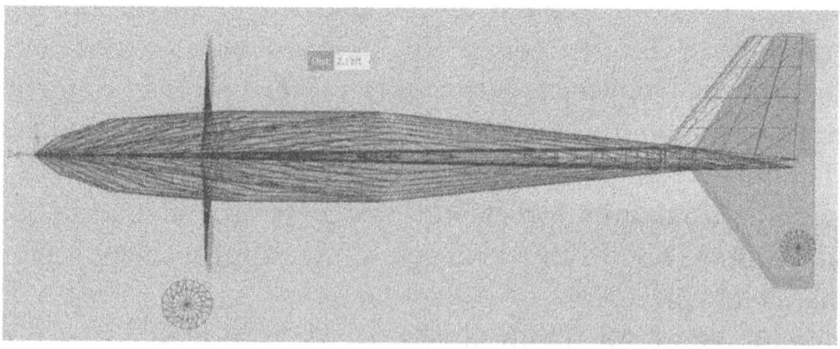

Fig. 5.28 Model of UA with elliptical wing, created for the X-Plain flight simulator: side view.

Fig. 5.29 General view of the model.

characteristics of engines. In addition, among the X-Plane there are super-critical flight regimes, which are especially important in this research.

Fig. 5.30 shows the model of the control system for the pitch channel. The block "Get data from XPlane 1" receives data on the pitch angle, then a comparison with the desired value occurs, then the result goes to the PID controller. The block "Send to XPlane" sends data to the flight simulator. On Fig. 5.31 screenshots of the flight simulation of UAs are presented.

5.16 Conclusion

In the process of studying the controllability of UA with elliptical wing, its own flight data collection and storage system was developed based on the modern high-performance microcontroller STM32f407. Flight data is not only stored on a flash-card, but also transmitted in real time via a separate radio channel. The onboard recording device also included a barometric altitude sensor, a GPS module, and a differential air pressure sensor.

For the development of hardware and software tools, a model-based approach based on Matlab/Simulink was used, which made it possible to simplify the development of both embedded and computer-used software. The use of this approach also allowed us to create a number of programs for analyzing flight data.

The flight tests of the prototype UA with an elliptical wing made it possible to determine the characteristic features of the behavior such as:

- stable flight at positive values of the angle of attack of the wing from 5^0 to 45^0;
- control damping along the roll channel relative to the pitch channel;

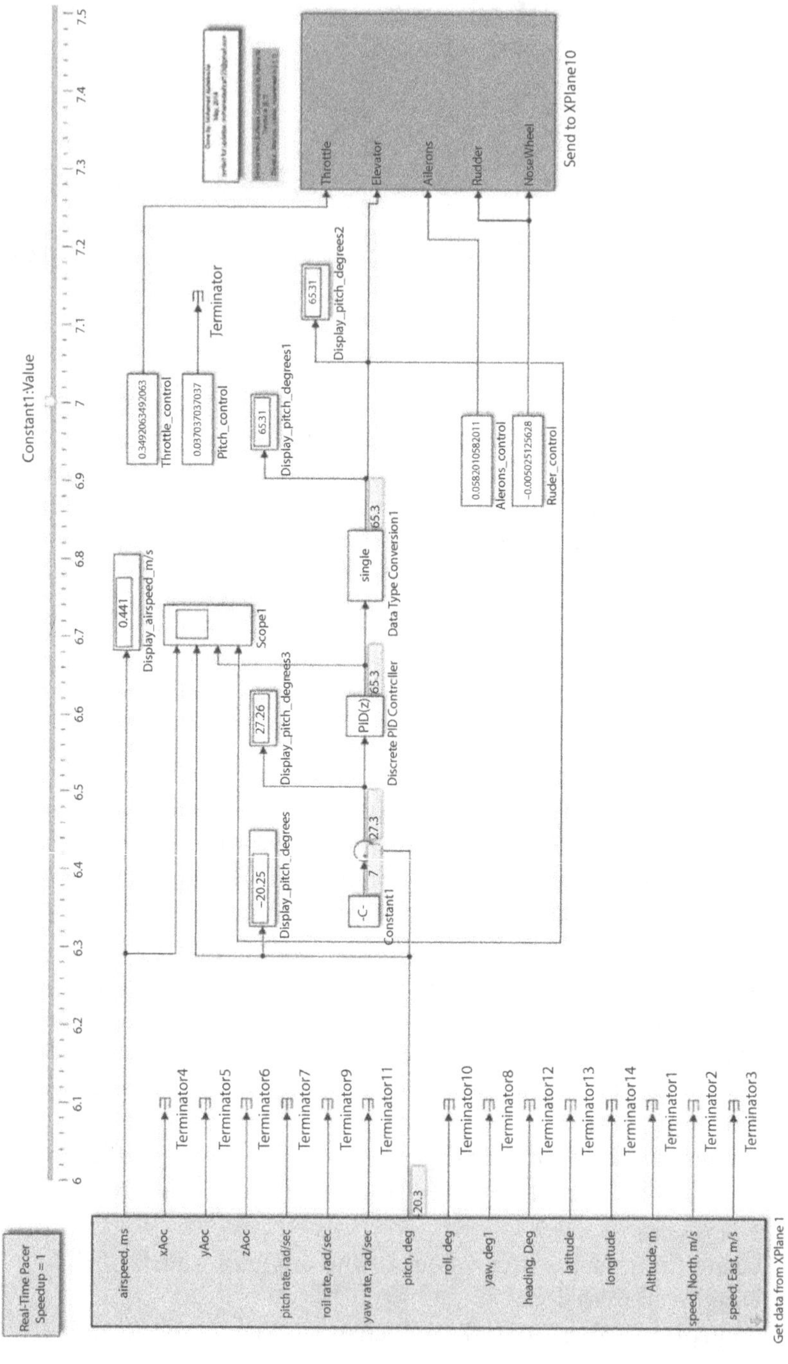

Fig. 5.30 Pitch angle control channel in Simulink.

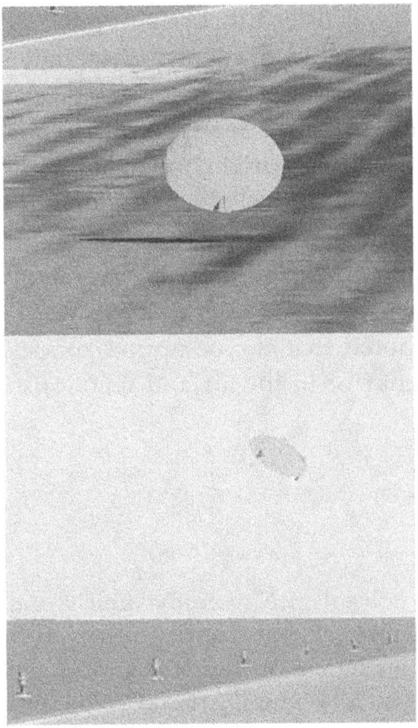

Fig. 5.31 Screenshots of UA flight simulation.

- a strong relation between UA centering and roll damping;
- it was revealed that with a decrease in thrust and non-interference in the control process, the aircraft lowers its nose, picks up speed and self-stabilizes;
- increased sensitivity of control over the pitch channel in the case of giving the rudders from themselves, with positive values of the angle of attack close to zero.

To solve the problem of identifying the features of UA control with the subsequent identification of its mathematical model, a program was created that implements the collection, storage, analysis and display of flight data in real (quasi-real) time. In the process of research, a mathematical model of the aircraft was selected and substantiated in an analytical form, but without additional studies of aerodynamic coefficients, it can only give a general idea of the behavior of the UA and does not take into account the specific features of the wing geometry.

To determine the characteristics of controllability in the pitch channel that is critical for this type of UA, it was applied using the black box method and as a result, transfer functions were obtained. Next, the mathematical model was adapted from experimental data using machine learning methodology. As a result, the presence of a model, with high accuracy, describing the control object, made it possible to correctly set up the regulator, thus synthesizing the desired control law, which takes into account the specific behavior of the control object under study.

Verification of the synthesized control law was carried out by computer simulation of the flight in the XPlane environment using the computer model of the UA with an elliptical wing specially created in the Plane Maker editor. It should be noted that the computer model really demonstrated similar features of behavior in the air and reactions to control actions, as was the prototype.

References

1. Trupkin, V.V. and Solyanik, P.N., Aerodynamic layout of a promising aircraft with a low-lengthening wing. *Aerosp. Eng. Technol.*, 3, 11, 65–68, 2004.
2. Grischanov, V.V., Malinin, A.V., Tarasenko, M.M., Investigation of the aerodynamic characteristics of spinning (fan) UAV GDP using numerical modeling in the software package FLOWVISION. *International Forum Engineering Systems - 2015 April 6–7, 2015 Abstracts of FlowVision Moscow Engineering Systems*, 2015.
3. Gorbunov, A.A. and Sinutin, S.A., Mathematical model of an unmanned aerial vehicle with an ellipse wing. *Eng. Bull. Don.*, 1, part 2, 2015.
4. Sinutin, S.A., Gorbunov, A.A., Gorbunova, E.B., Slow-flying UAV with elliptical wing. Control features. *Izvestiya SFedU. Eng. Sci.*, 57–62, 2016.
5. Gorbunov, A.A. and Gorbunova, E.B., Obtaining a mathematical model of a UAV with an elliptical wing using the black box method. *Materials of the International scientific-practical conference of young scientists of the countries BRIKS –Rostov-on-Don*, Publisher SFEDU, pp. 60–64, 2015.
6. Gorbunov, A.A. and Gorbunova, E.B., Development of a control system for an unmanned aerial vehicle with an elliptical wing. *Collection of scientific articles of the All-Russian Scientific Conference of Young Scientists, Postgraduates and Students "System Engineering-2015"*, Publisher SFedU, Taganrog, pp. 48–54, 2015.
7. Smith, P.F., Prabhu, S.M., Friedman, J.H., The MathWorks, Inc., *Best Practices for Establishing a Model-Based Design Culture*, SAE Paper 2007–01–0777.
8. Kerry, G., Vinod, R., Sasaki, G., Dillaber, E., The MathWorks, Inc., *Large Scale Modeling for Embedded Applications*, SAE Paper 2010-01-0938.
9. Aimagin Co, LTD, *Waijung Blockset Manual*, 2013, //Electronics paper http://waijung.aimagin.com/.

10. Lewis, F.L., Stevens, B.L., Johnson, E.N., *Aircraft Control and Simulation. Third Edition. Dynamics, Controls Design, and Autonomous Systems*, 768p, Wiley, 2016.
11. Dunn, T., *Designing a New Breed of Flying Disc*, 2014, URL: http://www.tested.com/tech/robots/460940-designing-new-breed-flying-disc.
12. Bodner, V.A., *Automatic control systems for aircraft engines*. M. Engineering, 248p, 1973.
13. Hafer, X. and Sachs, G., *Senkrechtstart technik. Flugmechanik, Aerodynamik, Antriebssysteme*, p. 376, Springer Verlag, Berlin Heidelberg, 1982.
14. Klein, V. and Morelli, E.A., *Aircraft system identification: theory and practice*, Published by American Institute of Aeronautics and Astronautics, Inc., ISBN: 1-56347-832-3 (alk. Paper 486), 2006.
15. Haykin, S., *Neural Networks: A Comprehensive Foundation*, p. 1104p, Translation from English - M.: Publishing house "Williams", 2006.

6

Technology of Geometric Modeling of Dynamic Objects and Processes of Virtual Environment for Aviation-Space Simulators Construction

Valeriy G. Lee

Southern Federal University, Engineering Technological Academy, Department of Engineering Graphics and Computer Design, Taganrog, Russia

Abstract

The sixth chapter considers the technology of planning and implementing geometric support for simulator-modeling computer complexes operating on the basis of the virtual reality environment. The main elements of geometric algorithms for rational discretization (polygonization) of complex technical objects described by curved lines and surfaces are given. The main criteria for the quality of modeling are static and dynamic realism of objects and the provision of real-time display. The following describes the software architecture of the simulator complex with visualization channels of the virtual environment and technical vision. The final part presents examples of the complex functioning when conducting training sessions in the form of visual rows of the virtual modeling space.

Keywords: Virtual reality, discretization, simulation, frame information capacity, curves and surfaces, manipulator, polygonization, training and modeling stand

6.1 Introduction

The scientific and technical field of creating computer systems for simulating the virtual reality environment (SVR) is very diverse and includes such

Email: livg48@mail.ru

Iftikhar B. Abbasov (ed.) Computer Modeling in the Aerospace Industry, (211–260)

areas as: software and hardware tools for processing graphic images and their visualization, technical tools for implementing relevant processes, geometric issues of building objects and transforming them into discrete form, etc. No less extensive is the list of SVR applications – from computer games and advertising to industrial design and training equipment.

Theoretical and applied scientific work in the field of simulator engineering is aimed at studying issues related to simulation computer modeling; for example, for ground preparation of cosmonauts for extravehicular activities of EVA on the international space station ISS [1]. The informational basis of such systems is the creation of interactive planning tools and visual monitoring of the cosmonaut's activity at EVA process [2]. These issues are related to improving the efficiency and safety of the EVA frameworks, as well as to the training and simulations of the robotic equipment RNS cosmonauts-operators.

The need for these studies is conditioned upon the difficulty of preparing astronauts to perform EVA tasks, due to a number of reasons, the main of which are the following:

- need to create training conditions as close as possible to the conditions of performing actual operations;
- need to work out abnormal situations in the EVA process;
- complexity of configuration, multimodular ISS;
- inability to train astronauts using real equipment in full and in the desired configuration;
- use of robotic maintenance systems;
- application of remote control technologies for robotic complexes, etc.

Simulator technologies have now been formed in the successfully developing branch of the global industry [20, 21, 22, 23].

The scale of the tasks related to the development of the SVR for space simulator is reflected by the image of a real ISS (Fig. 6.1) and one of the types of equipment on its surface – the RTS of the European Aerospace Agency of the ERA type (Fig. 6.2).

You can also note the widespread use of modern simulators in the training of pilots. The work [3] describes the development of a highly efficient simulator, which is widely used in the modern aerospace industry based on a virtual verification system. The instruction set architecture feature is studied; a high-performance graphics processor is used to improve performance. The work [4] is devoted to the

Fig. 6.1 Image of the international space station.

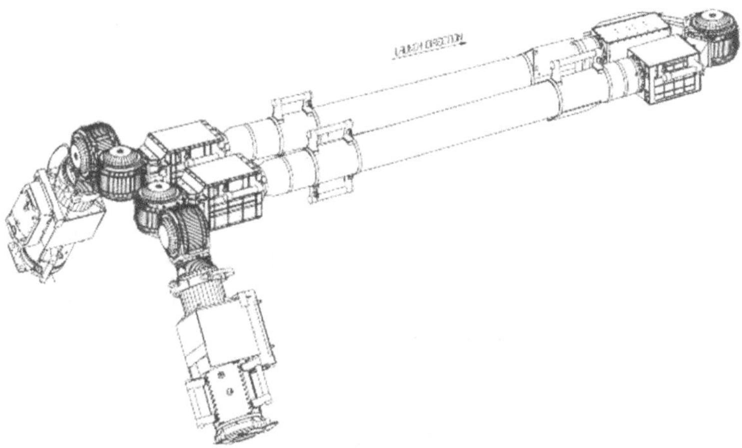

Fig. 6.2 Space station equipment.

development of visualization acceleration methods for the simulator. The adaptive selection scheme is implemented on the basis of the OSG Windows library; the acceptable rendering speed is increased without attracting additional computational costs. The work [5] explores the task of creating a database for unforeseen spacecraft operation scenarios. Contingencies would include terminating I thrust transmission of the explosion and debris, vector control loss, reversals, etc. The risk analysis of the impact of emergency scenarios on the life expectancy of unguided suborbital carrier rockets was carried out.

Work on the creation of hardware and software for the formation of the environment in the form of SVR and modeling the actions of the cosmonaut operator when working with RTS are intensively carried out in Russia and abroad. The components of visual modeling systems are developed using virtual reality tools (a system for creating three-dimensional images, a system for creating surround sound, a system for simulating interaction with objects of virtual reality, including touching objects, grabbing, throwing, etc.), virtual reality helmets, electronic gloves, mobile platforms (3-6 degrees of freedom), systems of formation of the panoramic image, holographic displays, etc. Complex simulators and system models are also developed.

Among the Russian developers of simulators for the aerospace industry, a number of organizations stand out, the results of which are:

- integrated simulator of the Russian segment of the international space station, providing training for astronauts in the interiors of the first two orbital modules of the Russian segment of the ISS;
- visualization system of the TEHNO procedural simulator for the formation of the visual environment of the virtual reality of the K-50 helicopter simulator;
- the training stand of the crews of the international space station using elements of the virtual reality of Atlas Aerospace;
- integrated simulator of the Russian ISS segment; training class on the program of the international space station (Yuri Gagarin Cosmonaut Training Center) and others.

The modern level of development of hardware and software, achievements in the field of ergonomics and psychology makes it possible to use models applied to information and advising systems that imitate the properties characteristic of human mental activity, as well as its perceptions, and, therefore, to consider the interaction between man and computer tasks in the paradigm of modeling systems of hybrid intelligence [2]. The quality of functioning of such systems largely depends on how the hardware, software, interface, etc., are consistent with human capabilities. At the same time, this is not about a person in general, but about a specific professional working in a particular subject area [6].

The proposed algorithms and programs for shaping the SVR were experimentally investigated in the development of simulation technologies for the activities of the cosmonaut-operator, and also tested in a number of articles published in foreign and domestic publications [7–9].

The specificity of the geometric problems of the synthesis of SVR in relation to the traditional areas of applications of applied geometry is reduced to the following main distinguishing features [7, 10, 11]:

- integration (complexity), arising from the need to model not only objects but also processes. Moreover, processes are understood as the dynamics of objects, and the dynamics of the entire virtual scene;
- a high degree of dependence of the applied synthesis technologies efficiency and tools on the hardware used in computational processing and visualization;
- the requirement to represent the entire set of information and control information flows in a discrete form, and the discretization parameters are due not only and not so much to the metric characteristics of the objects, but to the time intervals determined by the hardware environment. In this case, the main factor of realism is the reality of the time scale.

The most acute problem of combining high realism of the displayed scenes with the required rendering speed in real time is faced by the developers of the SVR in the aviation and space industries, due to the high technical complexity of the simulated objects and systems. Modern requirements for the visual component of SVR involve the formation of a full-color image with a high, real-time, rate of at least 22 frames/sec. with photorealism.

Figure 6.3 illustrates the structure of the main factors influencing the degree of visual and dynamic realism of virtual objects. It should be noted that at the present stage of development of computer technologies in the field of SVR synthesis, analytical methods of image synthesis are the most developed.

The main geometric and technological methods for improving the SVR realism:

- minimization of the discrete representation of curvilinear objects;
- optimization of discretization nodes placement;
- the use of detail levels of objects depending on the distance of observation;
- constructing curves directly in the modeling space;

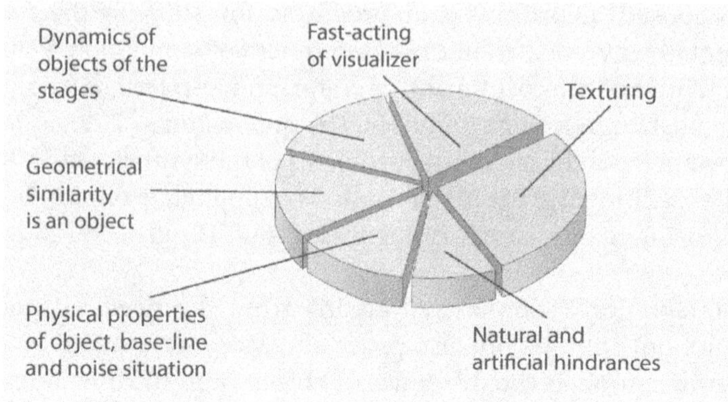

Fig. 6.3 The main factors that determine the realism of objects.

- fast algorithms for sorting and rendering polygons and textures;
- fast algorithms for implementing the dynamics of the behavior of objects;
- adaptability of geometric algorithms to parallel computing processes.

Analysis of the entire set of tasks related to the synthesis of the SVR allows us to single out the following main geometric problems [7]:

- minimization of geometric information on the objects of complex nature, provided that they provide sufficient realism of their computer virtual images;
- real-time display of dynamic processes in SVR, including the dynamics of changes in the shape and position of modeling objects;
- the development of the concept of end-to-end sampling of geometric and graphic information, ensuring maximum automation of all stages of its preparation, processing and storage – the so-called total systems of discrete geometric modeling.

6.2 Methods of Applied Geometry in Solving Problems of Simulation Modeling in SVR

Geometric and dynamic realism of objects and processes of SVR is largely determined by the accuracy of the discrete representation of objects and the speed of rendering graphic information. Despite the constant increase in the computational and graphical powers of modern computer graphics, the influence of the size of graphic data streams on the provision of real-time modeling and visualization of SVR remains noticeable.

As a rule, virtual objects are estimated by the number of polygons, which in turn is derivative from the dimension of discrete frames of curves and surfaces. Obviously, it is necessary to strive for optimization (in our case, minimization) of these dimensions.

Structuring the problem of ensuring the realism of the visual component of the SVR with the aim of formulating and setting specific tasks in terms of applied geometry is possible. However, to increase the efficiency of the expected solutions, it is necessary to formulate and accept the basic hypothesis of discrete geometric modeling – the existence of not only the rational from an engineering point of view, but also a discrete representation optimal in the absolute geometric sense of curvilinear forms [12].

6.2.1 Optimum Discretization of Curved Lines

In recent decades, the classical technological chain in the design, presentation, and reproduction of objects of complex geometric nature has been formed in applied geometry. In a generalized form, the sequence of operations is schematically depicted in Figure 6.4. The diagram as a whole reflects the technology of processing geometric information for complex technical surfaces. An important distinctive feature of this technology is

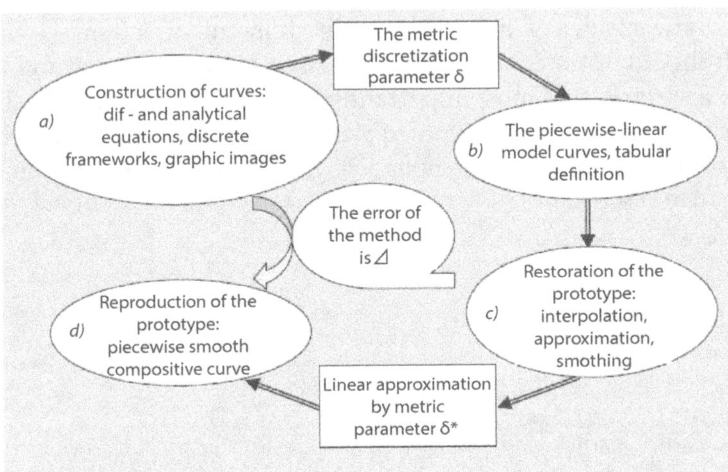

Fig. 6.4 Optimum discretization of curved lines.

the use of metric parameters δ and δ*, determining the quality of approximation and reproduction of geometric prototypes. Such a scheme is justified if high demands are placed on the objects to be processed on the accuracy of design and reproduction (production of precision engineering products, instrument making, etc.).

For the SVR modeling, another factor comes to the fore – the degree of realism of virtual objects and scenes. Analysis of researches in the field of model synthesis allows us to schematically reflect the structure of the main factors ensuring the realism of SVR. The following main directions of geometric studies can be formulated [7]:

- minimization of the discrete representation of curvilinear objects;
- optimization of discretization nodes placement;
- the use of detail levels of objects depending on the distance of observation;
- constructing spatial curves directly in the modeling space;
- fast algorithms for sorting and rendering polygons and textures;
- adaptability of geometric algorithms to parallel computing processes.

The problem of optimizing the discrete representation of curves is as follows. Classical and metrically oriented discretization (for example, by the deflection arrow δ) depending on the chosen direction, in the general case gives different results (Fig. 6.5). The reason for the discrepancy between the results of discretization in different directions is obvious, while it should be noted that δ* and δ** also obtained unequal and redundant in accuracy. And most importantly – the nodes *1* and *2* in both cases are not rational, which can be noted even visually. It can be argued that the nodes of such frameworks are not equal to each other in their geometrical informative significance.

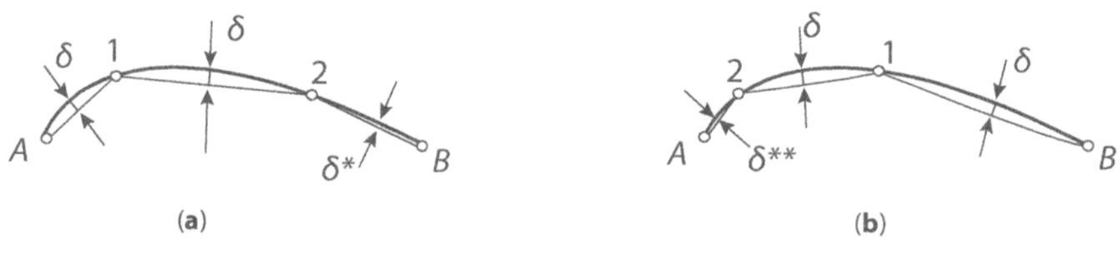

Fig. 6.5 Results of discretization.

At the polygonization of virtual objects, it is necessary to solve this problem in a different formulation. Namely, to place the discretization nodes at a predetermined number of them. Naturally, with the traditional approach, the problem becomes nonlinear and computationally difficult to solve, and in the case of a spatial curve, it is generally unsolvable.

At finding minimizing discretization, it is first of all necessary to strive to preserve, to the maximum extent, the geometric properties. Therefore, as criteria for the proximity of two curves, we should consider those variants that have a differential-geometric meaning. We will rely on the concept of isogeometric approximation, when the measure of closeness of partitioning intervals is understood to mean the degree of deviation of the integral set of differential characteristics of the maximum permissible order. An isogeometric approximation is introduced for functions of one and two variables on the basis of considering the set of points of change of the sign of derivatives. To approximate the function $y_1(x)$ by function $y_2(x)$ this is equivalent to using the variance characteristics

$$h_i(y_1, y_2) = \left| \frac{d^i y_1(x)}{dx^i} - \frac{d^i y_2(x)}{dx^i} \right|; i = 0, 1, 2, \dots, N.$$

Let us specify the criteria of proximity – h_i, having set the following requirements:

- the "proximity" characteristic of two objects must be effectively computable;
- the set of characteristics should be complete, that is, include a set of parameters sufficient for the application;
- the measure of deviation for each characteristic should have a fairly clear, practically applicable character.

It is known that, up to movement, the spatial curve is determined by curvature and torsion, and the surface is determined by the coefficients of the first and second quadratic forms; however, the direct use of these parameters as characteristics of proximity of curves or surfaces is problematic due to computational complexity and lack of clarity.

The task of finding the optimal discrete frame of a curve can be formulated in two versions:

Suppose we know some method of recovery curve, which allows for a given set of points $T_1, T_2, \dots, T_N$ build some interpolating curve f_2, satisfying the chosen criterion of proximity with the original prototype. Is there an extremum (in the sense of minimum) of the number *N*? The enhanced version of the problem is the addition of the question of the solution uniqueness.

On the basis of these two problems, it is easy to formulate the general problem of optimality of point frames of curves from the standpoint of applied geometry, where the criteria for the proximity of graphic images and prototypes are most important. Variational calculus suggests that for a given *N* for an arc of a regular curve there is an optimal allocation of discretization nodes, but an analytical (or numerical) solution of such a problem reduces to integrating some minimizing functional.

The task of approaching an arc by a regular and monotonic curve with a broken line in general form reduces to the following problem. Among the curves $y = f(x)$, belonging to the class $C^{(1)}$ on the interval $[a, b]$ and satisfying the condition $y(a) = m$, $y(b) = n$, find the one that implements the minimum functionality

$$I = \int_a^b F(x, y, y')dx.$$

From the variational calculus it is known that for an arc of any regular curve $y = f(x)$, we can always find such a broken line *P*, passing through the points $A(a, m)$ and $B(b, n)$, which will differ from the curve arbitrarily little as a deflection *d* and the angular coefficient of the tangent at any point of the interval. In other words, for a broken line *P*, functional value $I\{P\}$ will be different from $I\{f(x)\}$ arbitrarily small.

Conversely, for any broken line *P*, passing through the points *A* and *B*, such a valid curve can be built $y = f(x)$, passing through the vertices of the broken line, which is the value of the functional $I\{f(x)\}$ will also differ arbitrarily from $I\{P\}$.

Suppose *g* was the exact lower limit of the functional values $I\{f(x)\}$ on the set of all admissible curves. Then the following could be stated:

1) there is such a sequence of broken lines $\{P_n\}$, passing through the points $A(a, m)$ и $B(b, n)$, что $\lim I\{P_n\}=g$, при $n \to \infty$;
2) for each broken line P_i, passing through the points *A* and *B*, inequality $I\{P_i\} \geq g$, is made, that is a number *g* is the exact lower bound of values $I\{P\}$.

Let a minimizing sequence of broken lines be constructed. If a solution to the problem in question exists, then we can expect that the constructed sequence of broken lines converges to this solution. But the last conclusion will be valid only if the minimizing sequence of broken lines converges to an admissible curve, that is, if there is a possibility of passing to the limit under the integral sign, that is, that

$$g = lim\, I\{P_n\} = I\{lim\, P_n\}, \text{ at } n \to \infty.$$

There is an obvious process of building a minimizing sequence of broken lines. Suppose P_n is a required broken line, having n vertices in points $A_1(x_1, y_1)$, $A_2(x_2, y_2)$,..., $A_n(x_n, y_n)$ and passing through the points A and B, and it is necessary to fulfill the condition that all nodes are located within the interval, that is, $a < x_1 < x_2 \ldots < x_n < b$.

Since on the interval $[x_i, x_{i+1}]$ broken line element can be written as

$$y = f(x) = y_i + \frac{y_{i+1} - y_i}{x_{i+1} - x_i}(x - x_i) = y_i + \frac{\Delta y_i}{\Delta x_i}(x - x_i),$$

then after the substitution of broken line P_n we get

$$\mathrm{I}\{\mathrm{P_n}\} = \sum_{i=0}^{n} \int_{x_i}^{x_{i=1}} \left(F\left(x, y_i + \frac{\Delta y_i}{\Delta x_i}(x - x_i), \frac{\Delta y_i}{\Delta x_i} \right) \right) dx \left(x_0 = a, x_{n+1} = b \right).$$

After integration we find $I\{P_n\}$ as a function $2n$ of variables. If the numbers are determined $x_1, y_1, x_2, y_2, \ldots, x_n, y_n$ to the extent that the functional $\mathrm{I}\{x_1, y_1, x_2, y_2, \ldots, x_n, y_n\}$ reaches a minimum, then the sequence of broken lines P_n will be minimizing one.

6.2.2 Curve Integral Model

In differential geometry, it is possible to identify the relationship of a curve with a graphical representation of the parameters defining its shape. For a spatial curve, by analogy with the natural equations of its task, you can put in a one-to-one correspondence a graphical representation of its full curvature, since, according to the Lamé's Theorem $K(s)^2 = k(s)^2 + \tau(s)^2$,

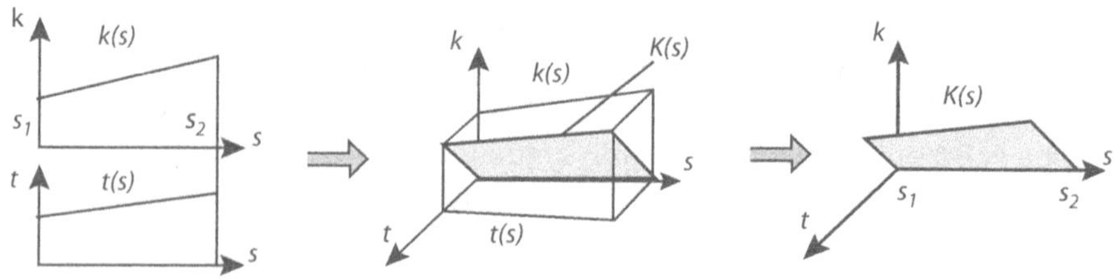

Fig. 6.6 Curve integral model.

where $k(s)$ is a curvature, $\tau(s)$ is torsion of a curve [13]. Then the following definition can be formulated.

Definition IMC integral curve model K, specified in natural parameterization on the interval $[s_1, s_2]$, called the compartment of a ruled surface with the plane of parallelism $Ok\tau$ and two directing lines: a segment $[s_1, s_2]$ and line $K(s)$ of the full curvature (Fig. 6.6).

6.2.3 Methods for Assessing the Information Capacity of Discrete Curve Frames

To implement a qualitative assessment of discrete skeletons of curves, we introduce the concept of their information capacity in terms of information theory.

For a fixed number N points, distributed in space end area D, the Crofton's Theorem, based on the Lebesgue measure $m(E)$ on distribution of a random number of points in a Euclidean space with a Poisson distribution with average $\lambda m(E)$, where λ – is a constant. For a finite number of points, the Lebesgue measure will be equal to $\{m(D)\}^2$, where D – is a subdomain of measurable space E [14].

Suppose we need to calculate the probability of an event that the figure R is formed by the N points, has the property, determined only by their mutual disposition, but not relative position R and D, invariance properties with respect to translation and rotation operations. As such properties assume the condition, that the distance between any two points does not exceed a certain predetermined value ρ. Then the probability of p, that meets the condition is equal to

$$p = \frac{m^*(R)}{\{m(D)\}^2},$$

where $m^*(R)$ is the Lebesgue measure of the set R corresponding to the imposed condition.

When constructing contour points, the condition is almost always satisfied that the indexing of the skeleton nodes is such that for any current point the previous and subsequent ones are the closest. Based on this, it can be confirmed, that for a finite set of unindexed points $\{R\}$, when imposing the condition of minimality of the distance when choosing a subsequent point, one can always obtain the law of indexation, which has the property of uniqueness and convergence, that is, the ordering of all points of a given set. Naturally, it is also necessary to specify that the procedures for implementing the ambiguity of said indexation (two or more equidistant points from the current point), namely, for the current point R_i the following is assigned such a point R_j, which has the closest point R_{j+1}, which, in turn, is the nearest to the point R_i. When sequentially joining indexed points, an unclosed spatial broken line without self-intersections will always be obtained.

The stability of a multipoint system shall be understood as its ability to respond to external influences. In our case, this is a change in the position of a point, or several points. The equilibrium state for some indexed set of points $\{A_i\}$ of the dimension N, is estimated by the value of the functional deviation U

$$U = u(\{x_i\}, \{y_i\}, \{z_i\})$$

from initial value U_0, when introducing into consideration some perturbations from the set $(\{\delta x_i\},\{\delta y_i\},\{\delta z_i\})$. The system is stable according to Liapunov, if in the presence of a perturbation, the magnitude UI stays close to the value $U(\{x_i\}_0,\{y_i\}_0,\{z_i\}_0)$ with a given rate ξ. The initial conditions for introducing the concept of stability of a point frame are as follows:

- two points in space uniquely define a straight line (as a special case of a curve); any number of points belonging to this straight line cannot improve its task, since it is already the limit. The information capacity of such a frame is maximal;
- the points located at the vertices of the tetrahedron (on the plane — at the vertices of an equilateral triangle) or a cube correspond to another limiting variant of the point set, since no indexing algorithm will allow us to construct a continuous smooth curve. The information capacity of such a framework is minimal (Figure 6.7).

Fig. 6.7 Location of points.

Let us introduce the concept of sustainability U of open point frame, which is estimated by a functional of the form [12]:

$$U = \max_{N}\left\{\frac{(B+C)}{L} - 1\right\},$$

where B – cut length $A_{i-1}A_{i+1} = \min_{K}\{A_kA_{k+1}\}$, for all $k = [1 : N]$;
C – cut length $A_iA_{i+1} = \min_{K}\{A_kA_{k+1}\}$, for all $k = [1 : N\text{-}1]$;
L – cut length $A_{i-1}A_{i+1}$, for all $k = [2 : N\text{-}1]$.

This functionality is built with two limitations: the number of points cannot be less than 3, and the value $(B + C)$ must exceed L. The failure of the last condition corresponds to the cases when the point configuration under study determines the object ambiguously, in other words, when $U < 0$ the system is unstable and the state of the object (the shape of the curve interpolating this point frame) does not reflect. For a steady state configuration, the value of the functional U must be between 0 to U_{max}:

$$U_{max} = \frac{2\left(\max_{l}\{A_lA_{l+1}\}\right)}{\min_{m}\{A_mA_{m+2}\}} - 1, \quad \text{for } m = [1 : N\text{-}2], \quad l = [1 : N\text{-}1].$$

Information capacity T of the point frame is defined as $T = \log_2(U)$ (bit).

6.2.4 Optimal Discretization Based on Integral Curve Model

Suppose $\gamma : I \to R^n$- is a regular curve. Parameter change (reparameterization) is a display $h : J \to I$, satisfying the following requirements:

- h- smoothly, that is all derivatives $h^{(i)}(s)$ exist for all $s \in J$;
- $h'(s) \neq 0$ at no $s \in J$; $s\hat{I}\ J$;
- $h(J) = I$,

h - bijection J on I, therefore, there is an inverse mapping $h^{-1} : J \to I$.

It is important to note that an affine scaling operation is a special case of reparameterization.

Display of h is a diffeomorphism in the categories of differential geometry. This concept can be associated with the concept of equivalent arcs of smooth curves.

Suggestion 1. Two arcs of a smooth curve are equivalent if there is a bijection – a mapping h, possessing the property of a diffeomorphism. Up to the operation of translation and scaling, the equivalence of arcs determines the equality of their spherical indicatrices of tangent and integral conic curvatures.

Corollary 1. The arc of a smooth curve and its equidistant are equivalent.

Corollary 2. Equivalent arcs of the curve of constant curvature and torsion are equal.

Corollary 3. Equivalent helical arcs are similar (equal to scaling accuracy).

Corollary 4. Arcs of a curve with a conic curvature equal to zero are equivalent if their integral total curvatures are equal.

IMC division into equal figures leads to the problem of nonlinear geometric programming, since only for cylindrical helixes and hyperbolic spirals do the results of dividing curvature and torsion graphs with equal area figures coincide with the result of dividing the integral curvature of the curve (Fig. 6.8; 6.9).

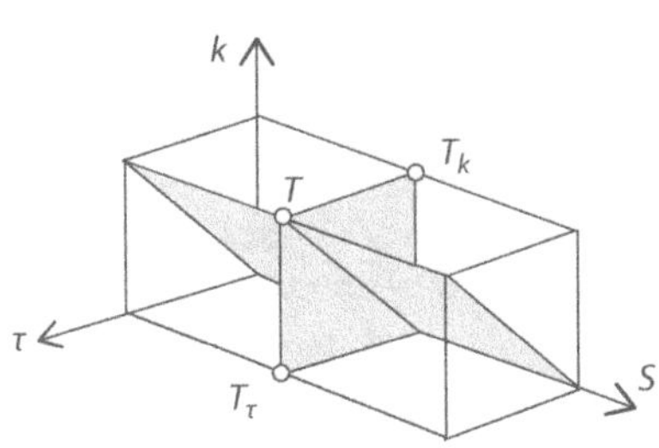

Fig. 6.8 Discretization methods.

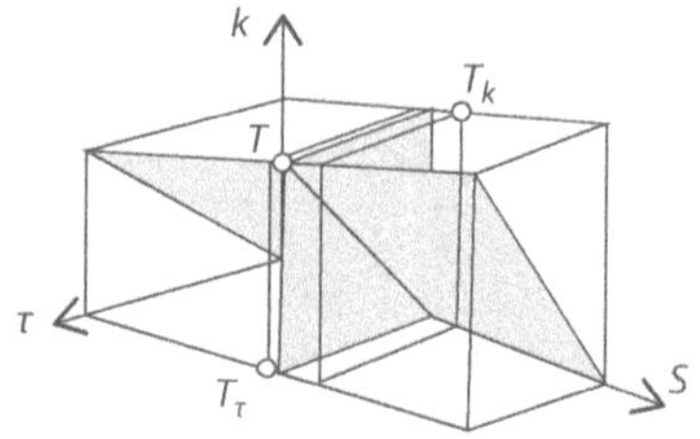

Fig. 6.9 Discretization methods.

In general, with an error of order $h^4(0)$ arcs of a smooth curve are equivalent if their integral cones of curvature and torsion are equal, that is, the integral total curvatures are equal [9].

The inscribed spatial broken line of a generalized helix, given in natural parametrization on the interval $[s_1, s_2]$, highlighting equivalent arcs on a curve, is determined by the expression

$$P_j = \frac{1}{j-1} \int_{s_1}^{s_2} K(s)\,ds,$$

where j - the dimension of the discrete frame;
$K(s)$ - full curvature of the curve;
P_j - integral (full) curvature of the extracted arc of the subinterval.

Geometric interpretation of the method: with an error $h^3(0)$ the next split point is located as the intersection point of the curve by a cone of rotation with an angle while forming P_j and the axis coinciding with the current tangent t_i (6.10).

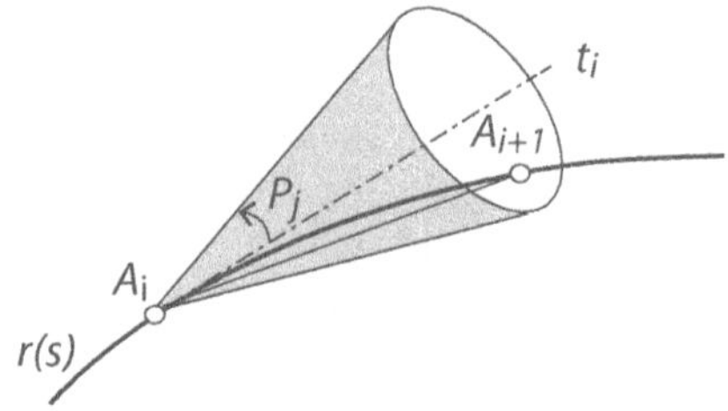

Fig. 6.10 Geometric interpretation.

Suggestion 2. For a given N discretization of R^* a broken line L is optimal if $T = \max_{R^*}\{T\}_N^j, \forall R_N^* \in L$.

Given the introduced concept of informativeness of the frame T we can formulate the signs of geometrically optimal discretization as follows.

Suggestion 3. The dotted frame of the curve is geometrically optimal if, for a given discretization parameter N he has the maximum amount of information T, and when removing any of the internal nodes, the information content of the frame is reduced by the same value.

Let us show the fundamental difference between the optimal discretization of a curve using the example of a flat monotone curve, which does not reduce the generality of the method.

Discretization by IMC has the property of additivity and uniqueness. Since for a given $N=3$ (in (Fig. 6.11) the integral model is divided into 3 trapeziums of equal area), as a result of which the discretization presented on Fig. 6.12 is shown.

The results of the analysis of the information capacity of the three variants of the framework of a single curve with the alternate removal of internal nodes are given in Table 6.1.

The experimental data show that the information content of the framework in Figure 6.12 is distributed more evenly between the nodes than in the frames of Figure 6.5.

To estimate the informativeness of the framework of a flat smooth curve, an equilateral triangle was used as a simplex (corresponding to zero information capacity). For the two-dimensional case (surface), a tetrahedron should be used as a simplex, as the minimum spatial configuration.

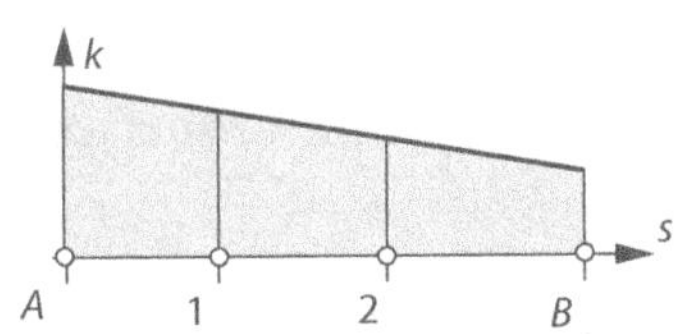

Fig. 6.11 Integral model.

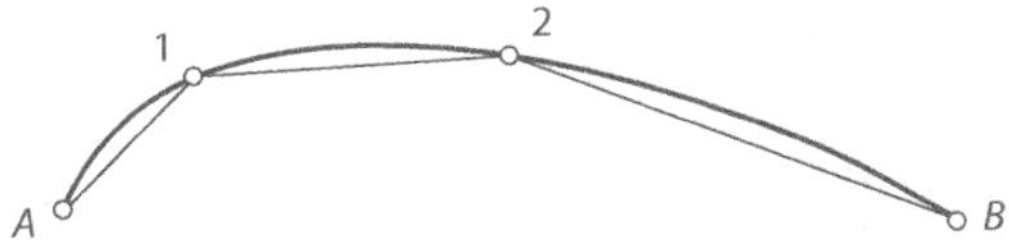

Fig. 6.12 Discretization of the curve.

Table 6.1 The information content of the three options for the frame of one curve.

Frame	Full information capacity of the frame	Information capacity without node 1	Information capacity without node 2	Average dispersion of nodes information capacity
Fig. 65a	3.38	1.45	2.12	19%
Fig. 6.5b	4.22	1.36	2.27	21%
Fig. 6.12	4.84	1.84	2.33	11%

In this case, the absolute will be the configuration when the vertex of the tetrahedron belongs to the base plane.

As an example, consider the frame $\{A_{m,n}\}$ of the compartment of surface dimension [M×N], namely: $m = 1,2,3; n = 1,\ldots, N$, where m is the number of frame lines; N is the number of points in the line. Then information capacity I of such a configuration is determined by the formula [15].

$$I = \log_2\left[\frac{2}{\left\{\dfrac{\sum_{m=2}^{M-1}\sum_{n=1}^{N-2}\left[S\left(A_n^{m-1}, A_{n+1}^{m-1}, A_n^{m+1}\right)\right] + S\left(A_{n+1}^{m-1}, A_{n+2}^{m-1}, A_{n+1}^{m+1}\right)}{\sum_{n=2}^{M-1}\sum_{n=1}^{N-2}\left[S\left(A_n^{m-1}, A_{n+1}^{m-1}, A_n^{m}\right)\right] + S\left(A_n^{m-1}, A_{n+1}^{m+1}, A_n^{m}\right) + S\left(A_{n+2}^{m-1}, A_n^{m+1}, A_n^{m}\right)}\right\}} - 1\right] \text{ (bit)},$$

Where S is area of a triangle with vertices A_n^m (n - frame node on the line m).

An experiment on a control curvilinear surface with an average value of the Gaussian curvature showed:

- dimension frame [3×13] has information capacity $I = 8{,}3$;
- thinned frame of the same dimension surface [3×7] – $I = 2{,}6$;
- thinned frame of the same dimension surface [3×3] – $I = 0{,}8$.

The last assessment reflects the fact of violation of the reliable display by the discrete frame of the geometric characteristics of the object.

From a methodological point of view, it can be stated that the geometric tools at the algorithmic level are basically formed, it allows to implement all the steps of the synthesis and visualization of the SVR (Fig. 6.13) fundamentally different from the traditional one (see Fig. 6.4). For the

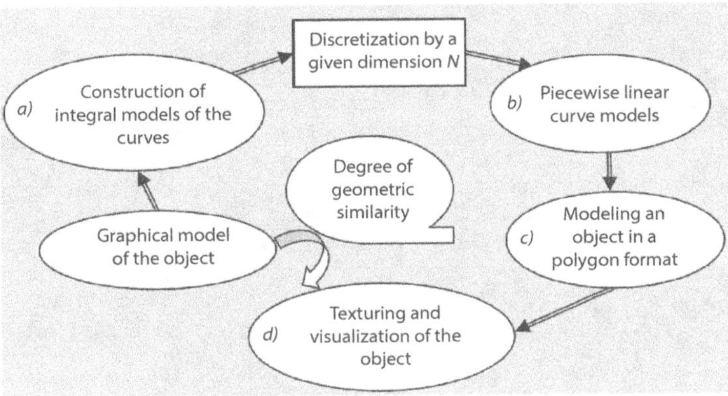

Fig. 6.13 Object synthesis technology.

computerization of the total polygonization technology, it is necessary to have the appropriate software and hardware in the form of a specialized graphic editor-modeler [7].

6.3 Purposes and Objectives of the Extravehicular Activity of the RTS Cosmonaut Operator on the ISS in Open Space, Technology of Computer Simulation in the Virtual Reality Environment

6.3.1 Extravehicular Activity of the RTS Cosmonaut Operator

The extravehicular activity (EVA) of an astronaut operator in outer space means different missions, where each begins with the exit from ISS through the airlock chamber, the implementation of the planned sequence of operations and ends with the return to the ISS (Figure 6.14). Each individual operation using the RTS is a sequence of astronaut movements (without moving its center of mass) and the mechanism controlled by it. Some of the missions allow execution in both normal and abnormal modes.

The EVA Operations Library contains the following main groups of scenarios for performing actions of the cosmonaut-operator.

1 group. Simple operations performed without the use of any instruments, tools, technical means:

- movement from the starting point located on the ISS surface to the target end point along a predetermined path;

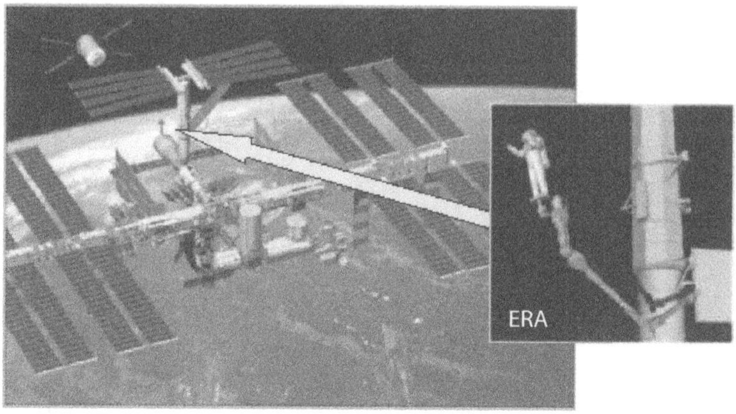

Fig. 6.14 Tasks of the cosmonaut-operator.

- visual inspection of the outer surface of the ISS, but for the case when the points of the trajectory are located at the station in order to search for a given object on it, malfunctions, etc.;
- practicing skills of maneuvering, targeting, assessing own orientation;

and etc.

2 group. Complex operations performed using the tools:

- delivery of the specified tool at hand (set of tools) to the specified point
- work as a given tool to achieve the goal;
- inspection of the surface of an ISS or its individual modules using one of the means of technical vision: a television camera, a movie camera, a thermal imager, a laser range finder, an infra-visor, etc.;

and etc.

3 group. Complex operations using robotic tools and control systems, for example, interactive control of a robotic manipulator.

Figure 6.15 presents a general view of the virtual simulation EVA space of the RTS cosmonaut-operator.

The managerial and control functions of the cosmonaut operator in relation to RTS are determined depending on the possible modes of his work [2]:

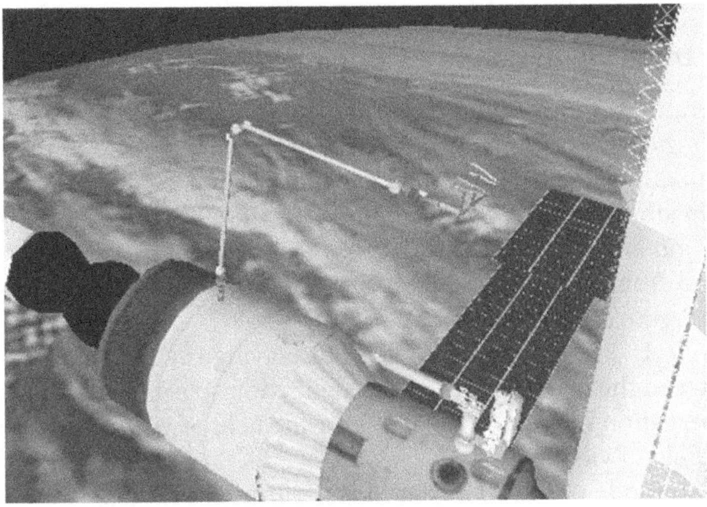

Fig. 6.15 General view of the virtual space.

- the supervisory mode, in which the system functions automatically in accordance with the training program recorded in the onboard control mini-computer (MC), the astronaut operator must assess the correct functioning of the system, record the execution of sequential operations and the program as a whole;
- interactive mode, which is a backup in relation to the supervisor; the operator, using the controls and the displayed information, controls with the help of virtual models of control panels of the RTSMC which informs the operator about dangerous situations and generates "instructions" for countering these situations;
- the auto-sequence mode, in which some operations are performed programmatically in an automatic mode, and intermediate operations must be performed either by the operator or by using another technical tool under the supervision of a training instructor;
- control mode by degrees of mobility, where the astronaut operator can set fixed values, for example, the angular velocities of turns in specified RTS hinges or the "step-by-step" control mode, in which the movements of the manipulator's working elements are specified in the form of fixed steps in degrees of mobility.

6.3.2 Technologies of Methodical and Hardware-Software Implementation of a Cosmonaut-Operator's Simulator

The operations of monitoring and tracking in the EVA process are carried out with the help of cameras installed both on the surface of the ISS and on the manipulator itself. In the virtual environment of a two-computer trainings simulator (TMS), in addition to these features, there are also observation modes from an arbitrary external point of the simulation space and from a point located on the cosmonaut's helmet (auto-tracking mode) (Figure 6.16).

Figure 6.17 presents an example of displaying the dynamic model of a helicopter in the working area of the simulation space – "auto tracking" in two observation modes: on the left is a wide field of view; on the right is a narrow field of view.

Manual mode of the RTS control is performed by an astronaut located in the basket of a mobile workplace or on the surface of the ISS using a fixed manipulator control panel located in the ISS in cases where automatic mode is impossible (abnormal situations) or dangerous. The remote control system provides on-line manual control of the RTS in the event that an astronaut operator is located on the ISS surface and must be fully functional and user-friendly.

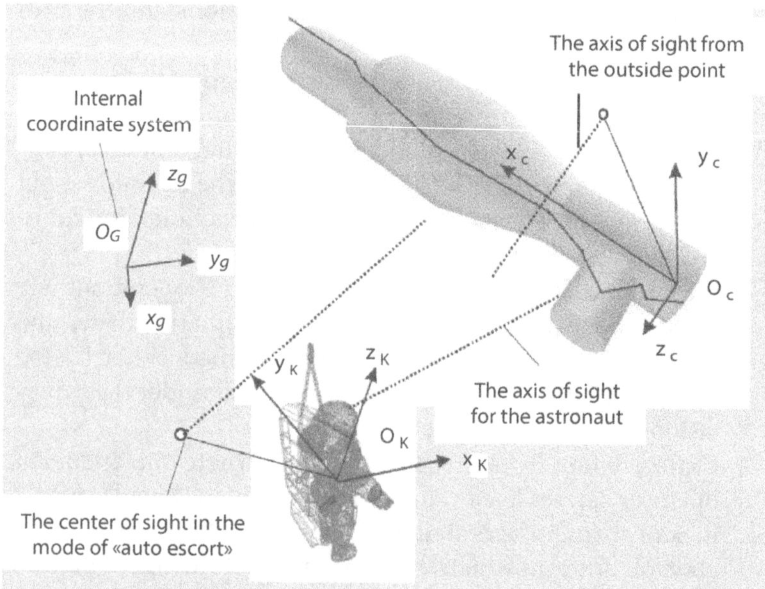

Fig. 6.16 Observation modes.

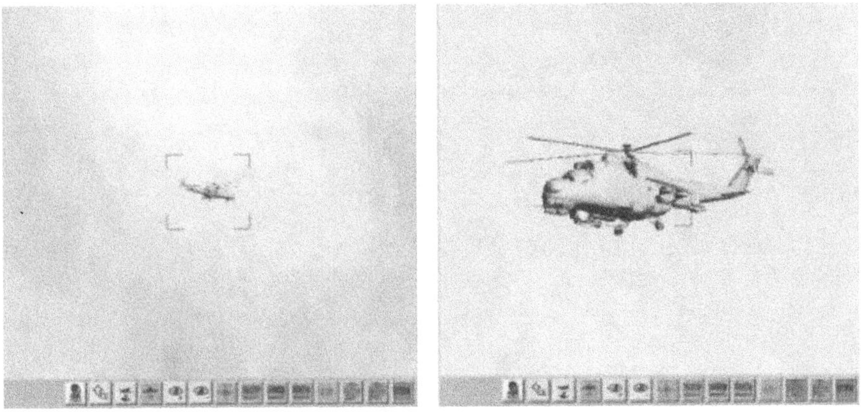

Fig. 6.17 Different fields of view.

In order to increase the reliability and safety of activities and the productivity of the work, the cosmonaut-operator should have duplicate channels of tracking, management and control of EVA. The most technologically advanced are remote technical means of autonomous implementation of EVA using.

A reliable and safe positioning method for ERA is the pre-simulation of EVA in virtual space. In this case, the virtual scene visible from an arbitrary (interactively controlled) outer point of space is displayed on the astronaut's stereo monitors (virtual reality helmet goggles).

When displaying the EVA work area, continuous analysis is performed to perform the necessary monitoring and evaluation of the quality of operations. In order to control the areas of safe movement of the working bodies of ERA, a method of current visualization of dimensional translucent spheres is applied (Fig. 6.18).

The task of target designation is to determine and define in three-dimensional space the location of the point that is the target of the movement of ERA. Targeting in the real environment EVA as an independent operation at any time of the exercise is carried out on the basis of the targeting process. Figure 6.19 presents one of the targeting options in a virtual environment.

Fig. 6.20 presents the main stages of target designation and controlled movement of the payload (cosmonaut in the basket), its subsequent spatial orientation and the output of the current spatial coordinates on the working display of the cosmonaut operator.

The orientation error in a spatial coordinate grid with a cell of 1 meter in size does not exceed 0.5 meter. The increase of orientation accuracy is

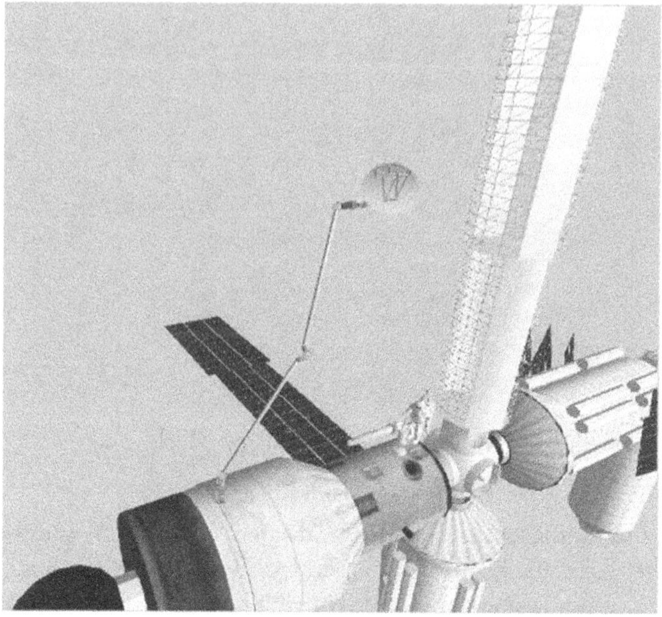

Fig. 6.18 Control of displacement zones.

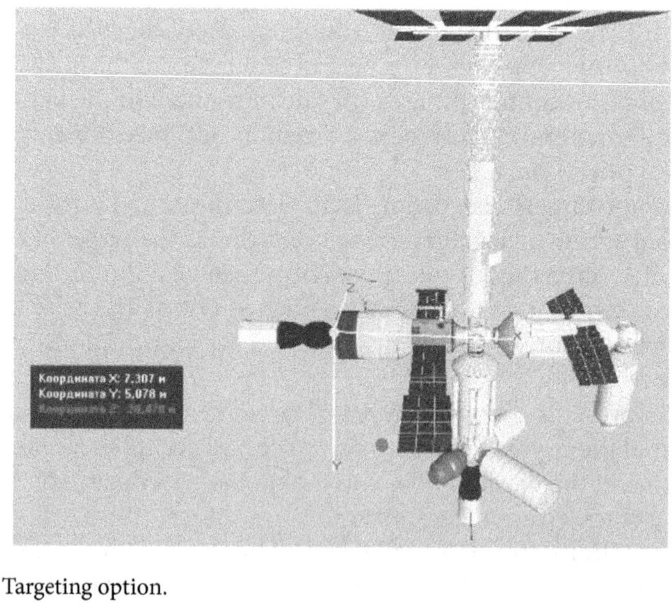

Fig. 6.19 Targeting option.

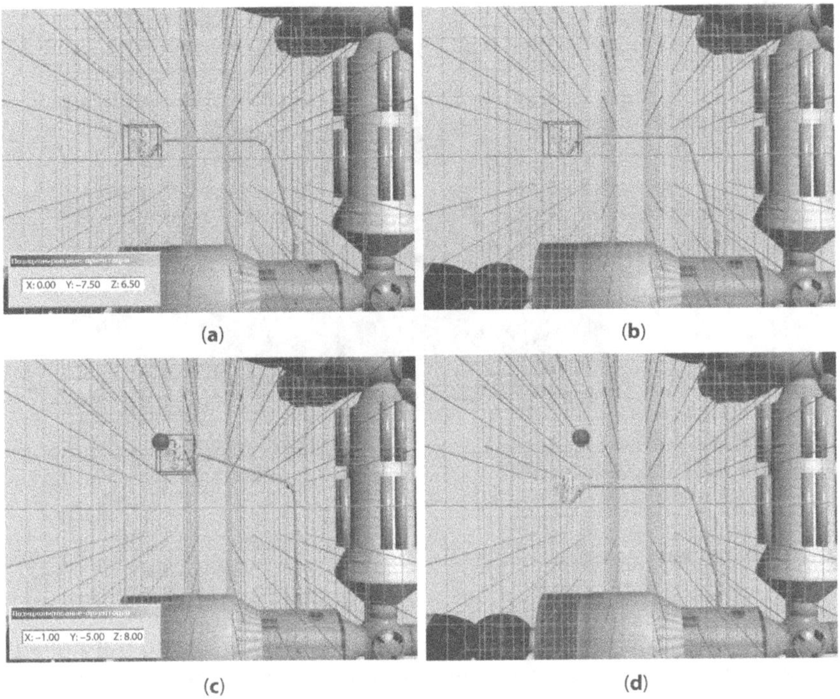

Fig. 6.20 Target designation stages.

possible, but not more than up to 0.25 meters, which is associated with the level of error of the three-dimensional graphic modeling of ERA kinematics.

In all states of the working window, you can work in one of two languages: Russian or English (Russian by default). Figure 6.21 shows a window in English version.

Managing the ERA manipulator model is possible in one of two options:

- using the virtual model of the autonomous control panel of the manipulator (see Fig. 6.21).
- keyboard of a personal computer on which the operator interface is installed.

6.3.3 Dynamic Virtual Model of the Manipulator

The virtual model of ERA manipulator is a graphical model of a real manipulator of a medium degree of detail (metric positioning accuracy – no more than 10 cm, angular – no more than 1 degree).

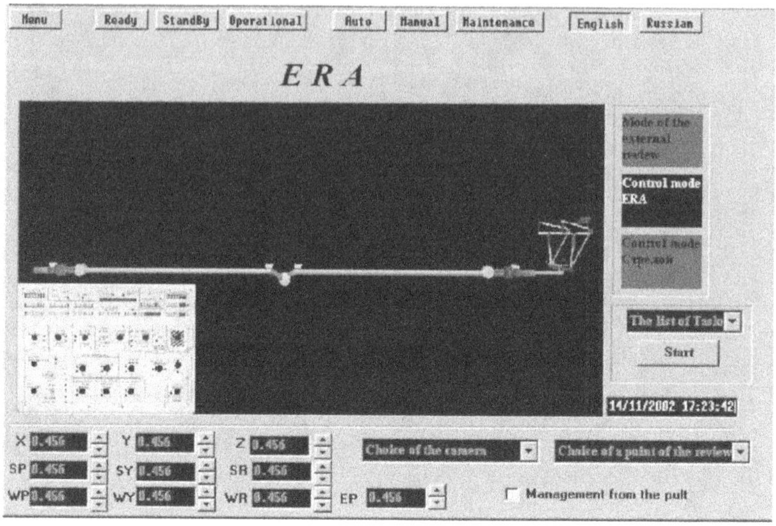

Fig. 6.21 Visualization window.

The virtual model takes into account the characteristics and functionality of a real manipulator, the implementation of which is carried out through:

- use of information expert system;
- the calculation of the angular velocities and working angles of the drives is carried out in the software package of mathematical modeling "*Euler 4*" [16].

After specifying the end point-target for moving the ERA handpiece in space, the software device analyzes the received coordinates and reports messages about the possibility/impossibility of moving the manipulator (the geometrical dimensions of the manipulator and working reach are taken into account, the maximum operating angles of the drives and the kinematics of the manipulator, 3-dimensional coordinates of the material, working and end point, the presence of obstacles in the path of the manipulator, etc.). Two modes of manipulator control are assumed: automatic and manual.

The ERA electromechanical manipulator consists of two identical sections (phalanx *A* and phalanx *B*), which are connected by the rotational hinge "elbow". The articulation model is presented in Figure 6.22.

Real ERA manipulator is a non-rigid construction, which significantly affects its dynamic characteristics when performing TVA operations, the overall length of the manipulator is 1180 cm.

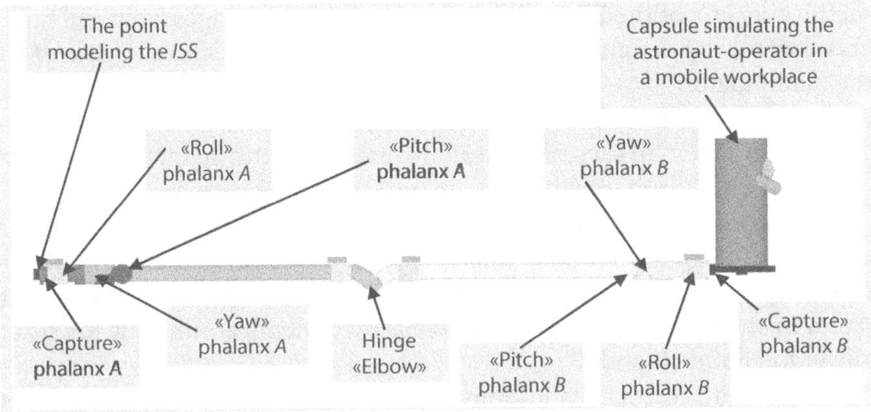

Fig. 6.22 Manipulator structure.

Dynamic realism of the virtual model of the manipulator is provided by the implementation of damped oscillatory movements of the end point depending on the current length of the manipulator and the angular velocity. Oscillations are modeled in the resulting plane of the total motion vector of the end point.

Fig. 6.23 shows the result of the display of the dynamics of ERA manipulator at start and braking after rotation at the elbow joint (angular velocity – 1 degree/sec., angle of rotation – 40 degree).

In the process of performing most missions and EVA operations using RTS, the working space is viewed using four television cameras mounted

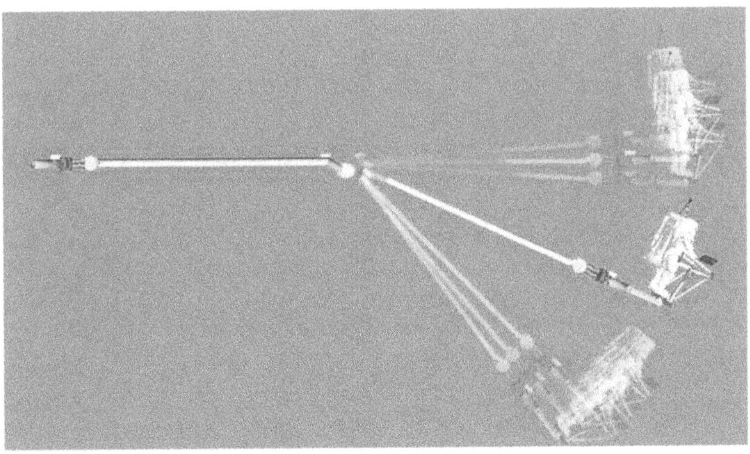

Fig. 6.23 Manipulator driving dynamics.

on the pharynx of the ERA manipulator (6.24). The images from the cameras control the dynamics of the movement of the structural elements of the manipulator.

Fig. 6.25 shows images obtained from each of the cameras in the process of a complex (in several planes) movements of the manipulator: *a*- *TV* camera 1; *b*- *TV* camera 2; *c*- *TV* camera 3; *d*- *TV* camera 4.

Fig. 6.26 shows the scheme of binding *ERA* to the local coordinate system $0_{ISS}xyz$ ISS centered at $0_{ISS}(0;0;0)$. At the initial time, the manipulator extends along the axis 0^1z and is rotated around this axis so that the point 0^5 lies in the plane 0^1yz, that there we no increments around the x-axis at any of the points 0^i and K^i.

Also in Fig. 6.26 an additional local coordinate system 0_{ERA} xyz is shown. In this case, the coordinate axes are collinear and aligned with the axes 0_{ISS} xyz.

Fig. 6.27 shows the transfers of the local coordinate system at specified points of rotation. Fig. 6.28 shows the length of manipulator joints and rotation angles of its joints, and the position of cameras. The notations given in these figures are:

- 0_{ISS} - the origin center of the local ISS coordinate system;
- 0^1 - the point of rigid binding of the manipulator to the ISS surface is characterized by the coordinate of the binding point $0^1(x^1 ; y^1 ; z^1)$;
- 0^2, 0^8 - phalanx roll points *A* and phalanges *B* accordingly, they have one degree of freedom - rotation around the axis 0^2 *y* and 0^8 *y*;

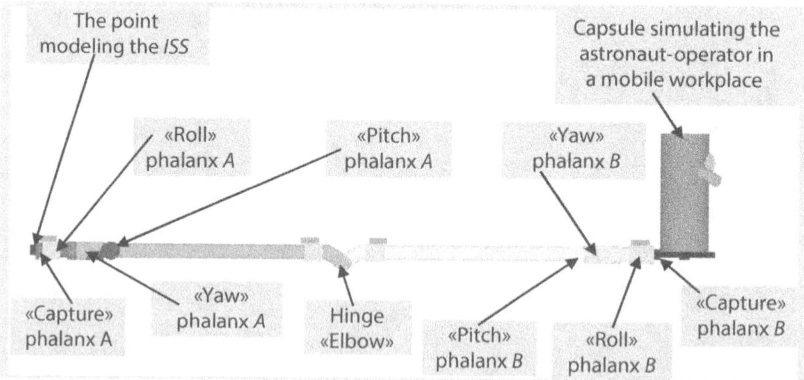

Fig. 6.24 Camera location.

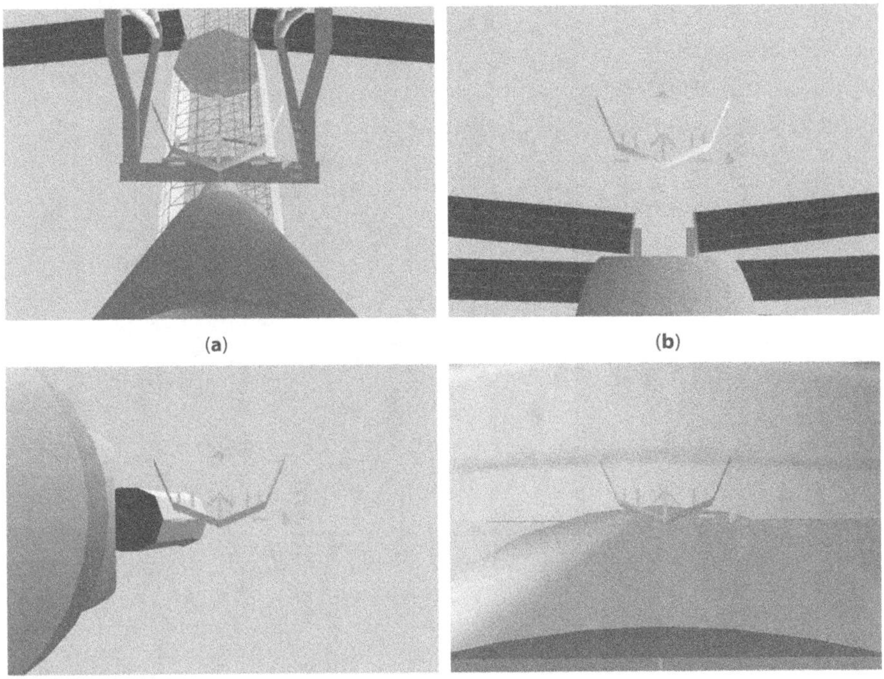

(a) (b) (c) (d)

Fig. 6.25 Images from different cameras.

- 0^3, 0^7 - yawing points of phalanx A and B phalanges accordingly have one degree of freedom - rotation around the axis 0^2y and 0^8y;
- 0^4, 0^6- Phalanx pitch points A and B phalanges have one degree of freedom - rotation around the axis 0^4x and 0^6x;
- 0^5 - the point of rotation of the "elbow" hinge, has one degree of freedom (rotation around the axis 0^5z).

6.3.4 Software Technologies for the Formation of Dynamic Models of the Editor-Modeler

For geometric design, editing or simplifying graphical objects and models in a graphical environment *World Up* modeler is used *World Up Modeler*.

The main commands of the language describing the model behavior of the scene (angles are given in degrees *(float)*, positions in meters *(float)*, time in seconds *(long)*:

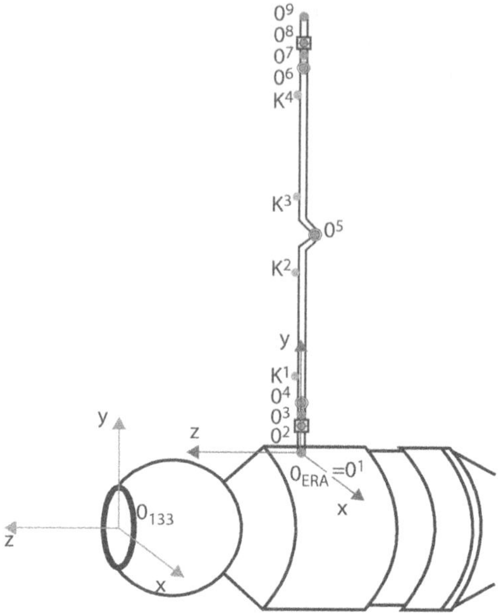

Fig. 6.26 Manipulator binding scheme.

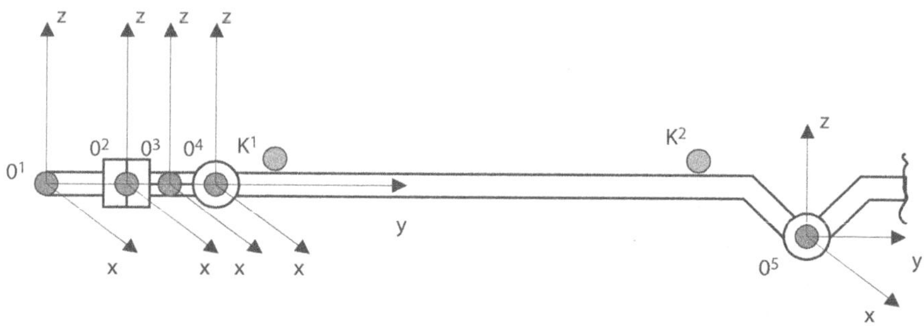

Fig. 6.27 Transferring the local coordinate system.

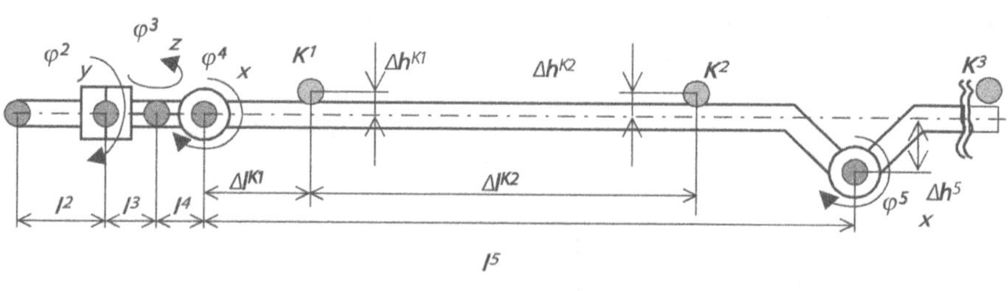

Fig. 6.28 Manipulator scheme.

- *SET X Y Z Th Ph Gm* - set the coordinates and angles of the object;
- *HIDE* - hide the object;
- *SETPOS X Y Z* - set only the position of the object;
- *SETANGLE Th Ph Gm* - set the angles of the object;
- *SETSPEED sX sY sZ* - set the speed in the vector representation;
- *SETROTATION sTH sPH sGM* - set the speed of turns in the corners;
- *MOVETO dX dY dZ* - move a relative distance;
- *ROTATETO dTH dGM dPH* - turn by relative angles;
- *ROTATE sTH sPH sGM Time* - rotate the object over time *Time,* with the specified speed;
- *MOVE sX sY sZ Time* - move the object over time *Time,* with the specified speed;
- *COMPLEXMOVE sX sY sZ sTH sPH sGM Time* - rotate and move the object over time *Time,* with the specified speed;
- *WAIT Time* - wait for the specified time (it makes sense after the command *HIDE*);
- *GOTO Operator Number* - go to the operator with the specified number. Creates run loops;
- *LINEMOVE Speed Time* - move at the specified speed in the current direction for the specified time;
- *ZOOM Number* - set the parameter of the lens of the observer;
- *END* - stop processing the object.

Fig. 6.29 shows the working window of the graphic editor-modeler in the process of designing a polygonal drone model.

6.4 Experimental Studies of the Functional Completeness of TMS Graphics and Software

6.4.1 Information and Functional Power of the TMS Visualizer

The information state of the visualizer is determined by three parameters (by analogy with a physical object in the three-dimensional coordinate system of Fig. 6.30.

- the amount of geometric information of the static SVR object model – axis *A*,

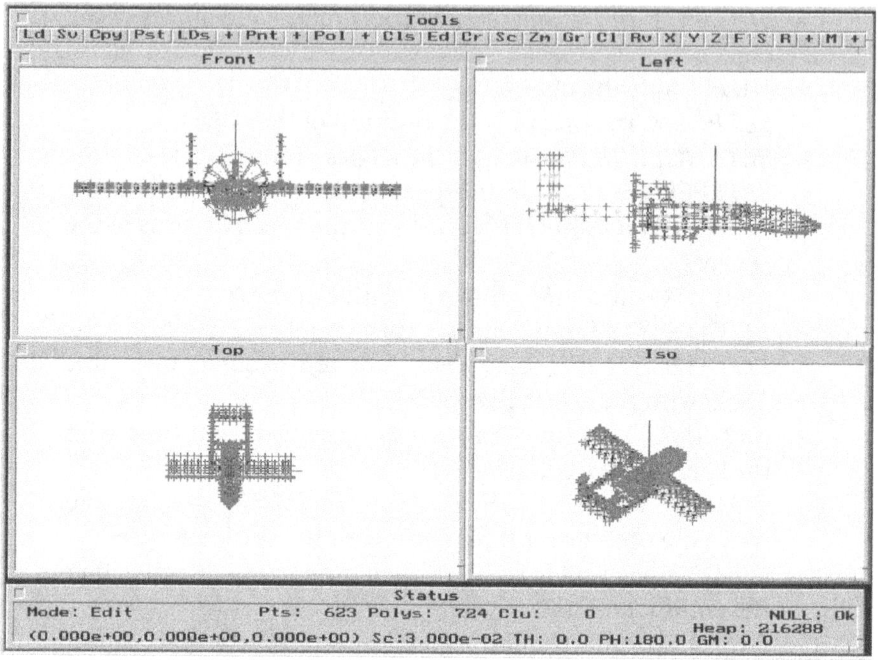

Fig. 6.29 Graphic editor window.

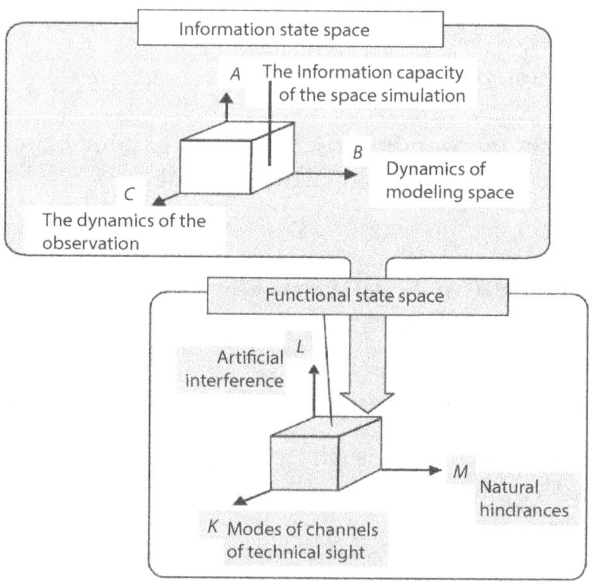

Fig. 6.30 Information state of the visualizer.

- dynamics of behavior (movement) of all moving models – axis *A*:
- the dynamics of the behavior (movement) of the observer – axis *C*.

Vertical axis *A* Displays the amount of geometric information contained in the static layout of the modeling space. Naturally, the amount of this type of information depends on the number of objects in the modeling space and their structural, geometric complexity.

The amount of information processed by the visualizer also depends on the dynamic characteristics of the objects and the observer. Moreover, this dependence will increase in proportion to the increase in the speed and complexity of the movements of the observed objects [17].

Specifically, the value of the function *A*- there is a total amount of geometric information of all elements of the modeling layout.

Informational state *I* the visualizer can be represented by the value of a view function $I = f_1\{A, B, C\}$, and A×B×C can be used as a quantitative characteristic – the integrated information power of the modeling space.

The functional space of the visualizer in terms of visualization tasks is determined mainly by three characteristics:

- the volume and complexity of the simulated artificial noise and effects: surface textures, object transformations, special effects of artificial lighting, smoke, fire, noise, made by means of technical vision, etc. – axis *L*.
- the volume and complexity of the simulated natural noise and effects: imitation of weather conditions, the results of natural lighting, etc. – axis *M*.
- type of observation, modes of operation of the optical system means of vision: resolution, focal length, zoom ratio, etc. – axis *K*.

By analogy with the informational assessment functional capacity *F*, the visualizer is defined as $F = f_2\{K, L, M\}$. The process of modeling can be represented as the process of "immersion" of space *I* into space *F*, which is reflected in the strategy of the chosen scheme of passing the geometric information and framing the synthesized images.

The outlined scheme of the formal representation of the information and functional spaces of the visualizer allows, on the one hand, to quantify its performance, and on the other hand, to reflect its specific state, in terms of the completeness of loading its computational and graphical resources, namely in the form of the functional $V = f_4\{I, F\}$.

Based on the developed specialized compiler (programming in the environment *WatcomC10*) and memory models *(float)*, when working in 32-bit *DOS*-processor safe mode (using the extender *DOS/4GW*), that is, ensuring linear addressing and access to all installed memory without the use of protocols *EMS* and *XMS*, use standard *VESA* and drivers *UNIVBE2/0*, which provides sufficient independence for the purity of the experiment on the type of video adapter) a set of experiments was carried out. The purpose of the experiments is to identify for the visualizer the threshold values of the information power of dynamic scenes at which there is a qualitative reduction in the level of realism SVR by criterion of speed.

Fig. 6.31 shows a fragment of a video sequence of experiments evaluating the performance of a visualizer with a virtual model of an aircraft (AU) (amphibious aircraft Бе-200).

Figures 6.32–34 in the form of three-dimensional graphs show separate results of the experiment, namely, for scenes containing one dynamic object (AC) with details – 1200 polygons) with different background and landscape powers (from 0 to 5000 polygons). Figure 6.35 shows a generalized graph of the performance of the visualizer for three types of detailing of a dynamic virtual model AC with a constant information power of the background and landscape – 2000 polygons. The experiments were carried out with the use of designer texture for AC and landscape, which, as the experiment showed, significantly slowed down the work of the visualizer.

6.4.2 An Example of a Simulator of a Typical Flight Mission at Solar Battery Installation

1. operation. The gripping mechanism of the ERA manipulator to connect the solar array.
2. operation. Using the remote control of the manipulator, deliver the solar cell unit to the docking zone on the farm of the habitable laboratory module.

Fig. 6.31 Video series of the virtual model.

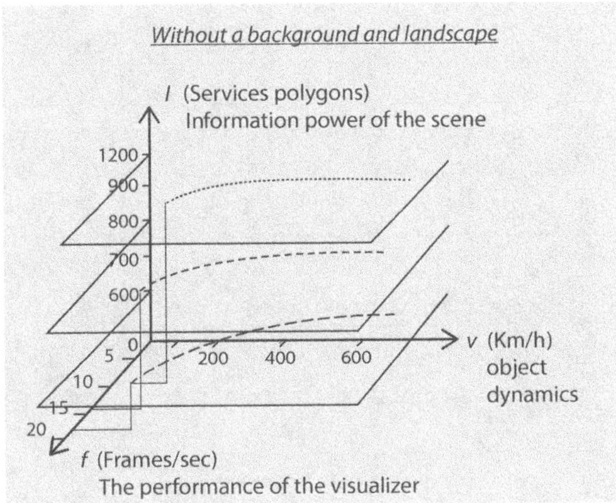

Fig. 6.32 Objects with different particularization.

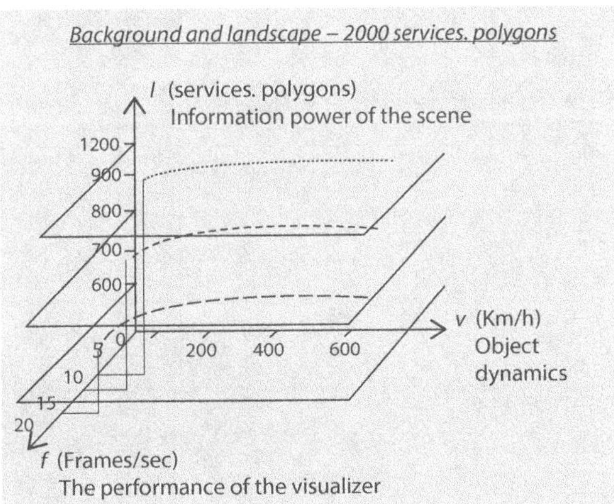

Fig. 6.33 Objects with a particularization of 2000.

3. operation. Dock the solar cell using the docking station on the ISS farm.
4. operation. Uncover solar panels.
5. operation. Return the working bodies of the ERA manipulator to its original state.

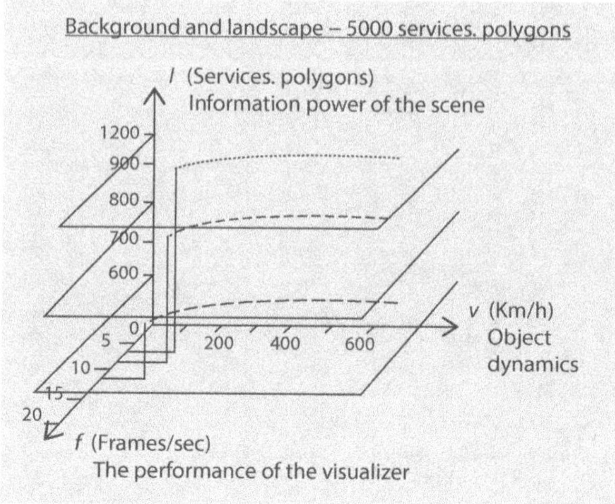

Fig. 6.34 Objects with a particularization of 5000.

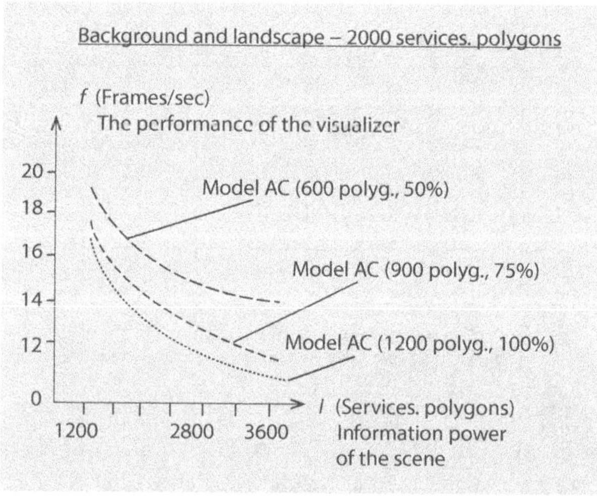

Fig. 6.35 Generalized schedule.

Fig. 6.36 shows keyframes of visualization of the scenario of the specified mission. Observation mode – from a virtual external fixed point of the modeling space. The observation distance from an external point is about 80 meters, and the field of view is wide.

Fig. 6.37 shows the frames of the operation of installing a solar battery using the *TV camera 4* optical sight mounted on *ERA*.

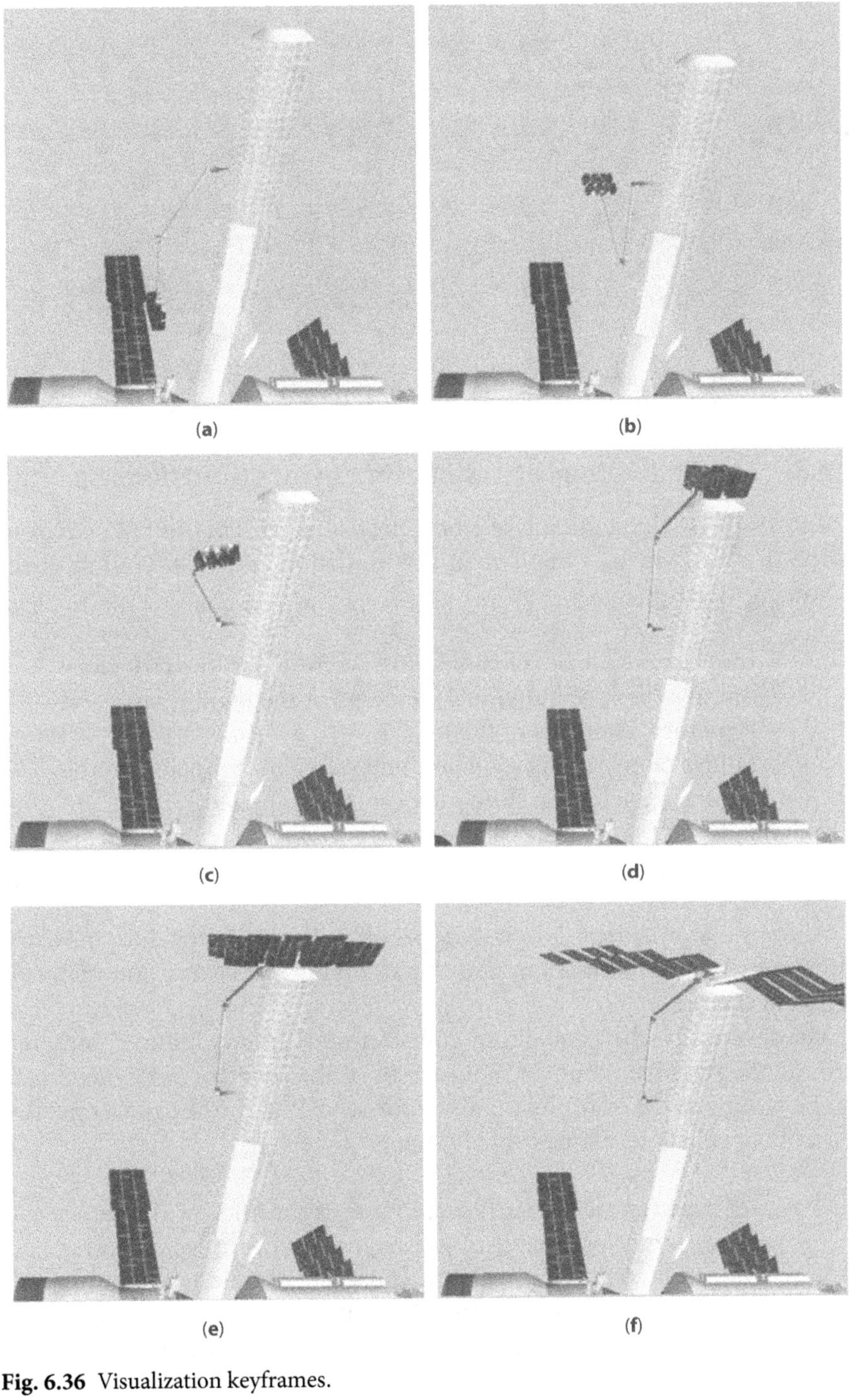

Fig. 6.36 Visualization keyframes.

(a) (b)

Fig. 6.37 Solar panel installation.

6.4.3 The Technology of Testing Emergency Situations

For software implementation of emergency situations in the TMS environment, it is necessary to implement the modeling and visualization of the following situations:

- the occurrence of collisions (intersections) of the trajectory of movement of individual elements of the manipulator and structural elements of the ISS.
- failure of any of the switches, buttons or scoreboards on the control panel when the control signal is applied;
- failure of any of the engines when giving a signal from the control panel.

The simulator instructor interface provides the ability to enter the listed failures or their combination, and the archiving subsystem automatically registers them.

Geometrically, the problem of collision modeling is reduced to determining the possible points of intersection of the trajectories of movement of the working bodies of the manipulator with ISS. At the same time, two variants of collisions are possible [5, 8]:

- intersection of ISS surface and a straight line. At the same time, one or both ends of the segment move in space along independent circles, which is caused by the design features of the ERA electromechanical manipulator.
- the intersection of the ISS surface (or its virtual model) and the trajectory of the point (the working point of the

manipulator or a sphere simulating the oscillation region of this point). In this case, the point moves in space along a complex trajectory, which is a superposition of curves lying on spheres, which is due to the design features of the ERA manipulator.

Table 6.2 presents several ERA-ISS collision options.

When implementing the above algorithm for solving the collision problem directly in a graphical environment visualizing the EVA virtual space of the cosmonaut operator, it is necessary to take into account the peculiarities of the presentation of all virtual objects in the used visualizer. Let us consider this problem on the example of a software-graphic solution of the problem of intersecting (touching) the ISS surface of one of the phalanges of the ERA manipulator.

During the operation of the manipulator, it is necessary to track the rotation of its elements in world coordinates. At the same time, the position of the current sighting axis "*B*" is fixed, and the objects "*A*" (elbow ERA) and "*C*" (phalanx ERA) can make independent movements (Figure 6.38).

In the modeling space, we have scene objects (polygons of graphic virtual objects) with known coordinates of the vertices of polygonal meshes with normals. Also known current coefficient display screen. Thus, it is possible to artificially cover the ISS model with a certain shell, increasing for this the display coefficient by some relative value proportional to the metric safety distance of the trajectory of the "*C*" object being developed.

After that, it is possible to automatically determine the distances from the analyzed points of the manipulator to the polygons of objects using the following procedure:

```
bool intersect_triangle_line(Cvector3 v1, // triangle tops
Cvector3 v2,
Cvector3 v3,
Cvector3 n,     // triangle normal
Cvector3 p1,    // first end of the segment
Cvector3 p2,    // second end of the segment
Cvector3 & pc,  // return point of intersection
// calculate the distance between the ends of the segment and
   the plane triangle
float r1 = n&(p1-v1);
float r2 = n&(p2-v1);
```

Table 6.2 Collision options ERA-ISS.

Collision option	Geometric collisions scheme	Mapping to SVR
Manipulator kick *ERA* in board ISS when turning in 4 hinge. The kick point is on the side of the ISS cylinder.	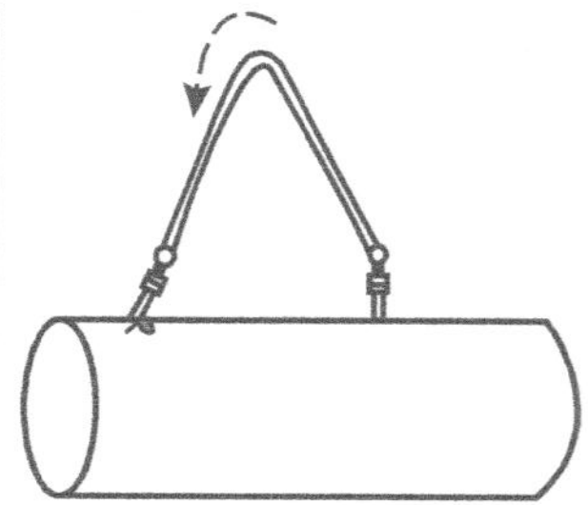	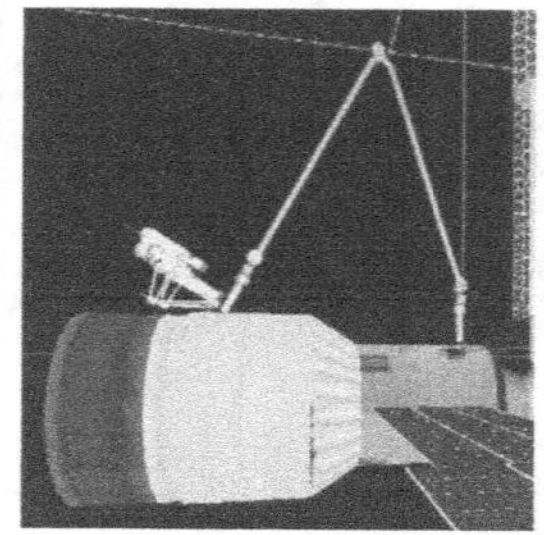
Manipulator kick *ERA* in board ISS when turning in 4 hinge. The kick point is on the end surface of the cylinder.	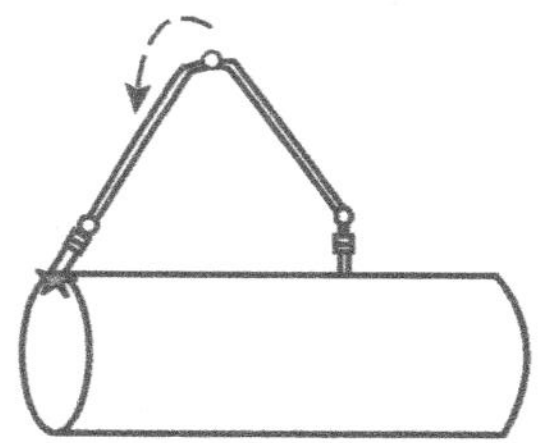	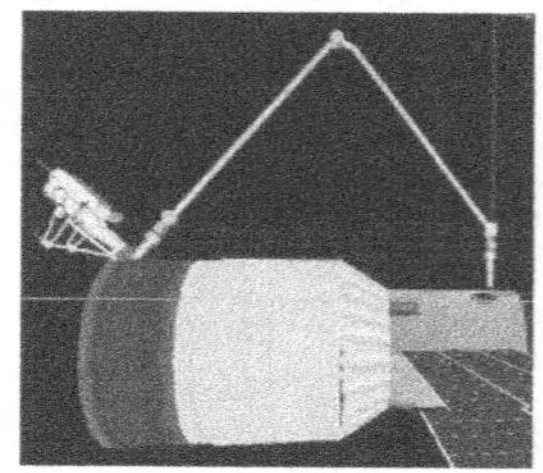

(*Continued*)

Table 6.2 Collision options ERA-ISS. (*Continued*)

Collision option	Geometric collisions scheme	Mapping to SVR
Manipulator grip punch *ERA* in board ISS when turning in 3 hinge. The kick point is on the edge of the base of the cylinder.	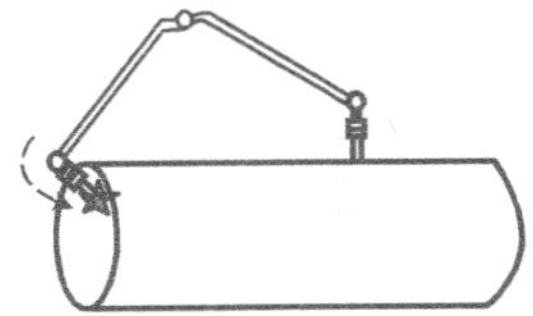	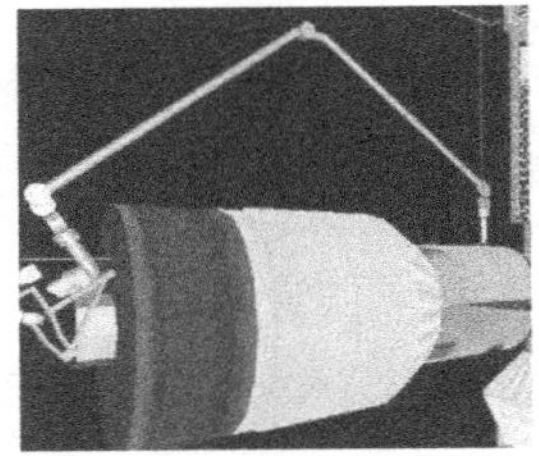
Capture kick *ERA* in board ISS when turning in 1 hinge. The kick point is on the lateral surface of the cylinder.	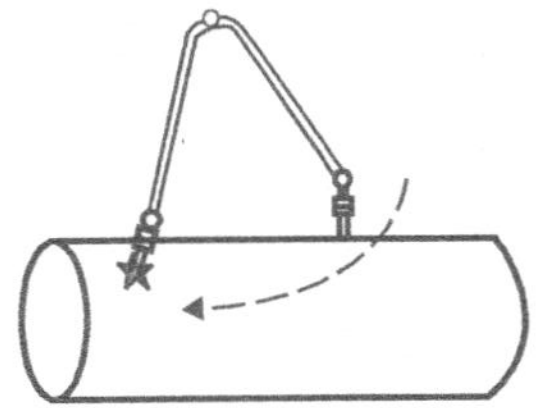	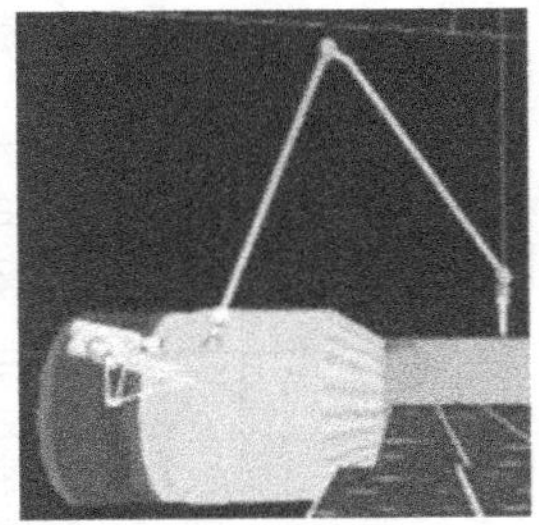
Manipulator phalanx kick *ERA* in board ISS when turning in 1 hinge. The kick point is on the lateral surface of the cylinder.	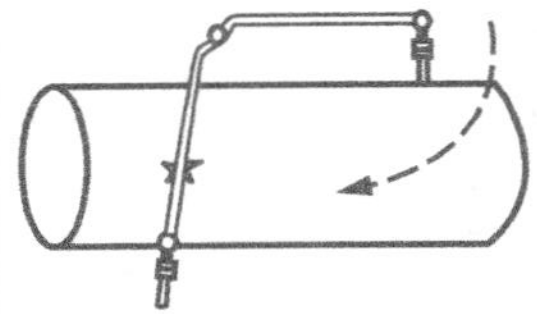	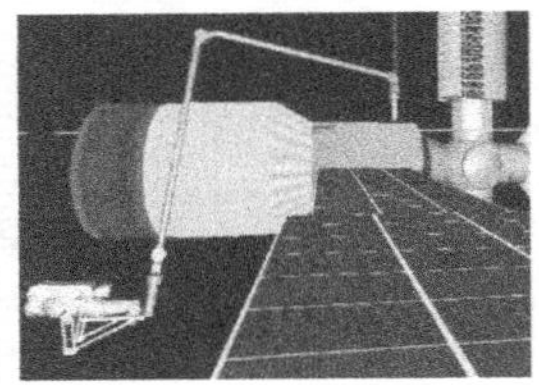

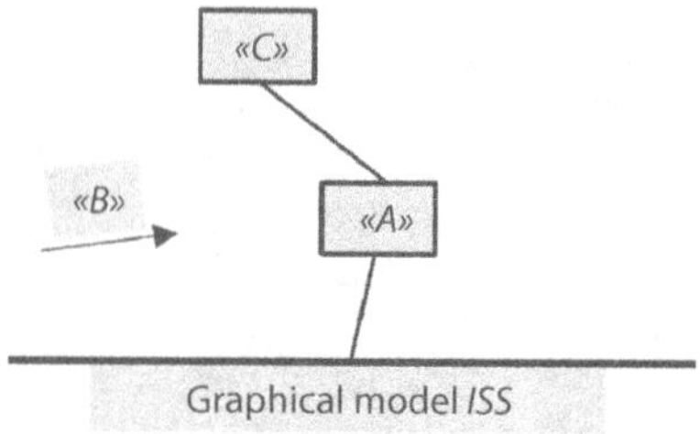

Fig. 6.38 Motion in the coordinate system.

```
// if both ends of the segment are on the same side of the plane, then the segment does not intersect the polygon
if (f1_sgn(r1) ==f1_sgn(r2) ) return FALSE;
// calculate the point of intersection of the segment with the polygon
Cvector3 ip = (p1 + ((p2 - p1)*(-r1 / (r2 - r1))));
// check the fact of finding the intersection point inside the triangle
if( (((v2 - v1)^(ip - v1)) & n) <=0) return FALSE;
if( (((v3 - v2)^(ip - v2)) & n) <=0) return FALSE;
if( (((v1 - v3)^(ip - v3)) & n) <-0) return FALSE;
pc = ip; return TRUE;
Dot Product u Cross Product,
int f1_sgn(const float &amp; k); //f1_sgn – function to calculate the sign of the result
{
if (k > 0) return 1;
if (k < 0) return - 1;
return 0;
}
```

To determine collisions, it is necessary to check the intersection of each ERA element with each polygon of other models. To increase the speed at the initial stage, the intersection of two most dimensional links with ISS polygons is checked. The intersection with the polygonal cylindrical model ISS is also analyzed. The radii of the cylinders of the polygonal cylindrical model are larger than the radius of the ISS surface since it is necessary to take into account the magnitude of the possible ERA oscillations during movement.

The problem of ensuring security during the execution of EVA is reduced, first of all, to the task of guaranteed prevention of any collisions of the working bodies of the ERA manipulator and the ISS surface. This can be achieved with the help of the technology of preliminary working out of the necessary trajectory of movement.

Fig. 6.39 shows the individual frames of the interactive search session for the safe movement of the manipulator to a given target point (highlighted in red). During the session, a protocol of control commands is generated that provide a safe trajectory.

In training sessions, only abnormal situations are worked out, where the possibility of performing EVA remains. The most characteristic of them are the occurrence of collisions and failures of one or several manipulator engines (no more than 4, with the exception of the "knee" engine).

The theoretical basis of work in abnormal conditions is the theory of rotations around the coordinate Cartesian axes, namely, the theory of Euler angles. In general, this theory reflects the possibility of replacing three independent rotations of the vector around the coordinate axes (u_1, u_2, u_3) two independent rotations (Euler angles). The result will be the same resultant position of the vector [18].

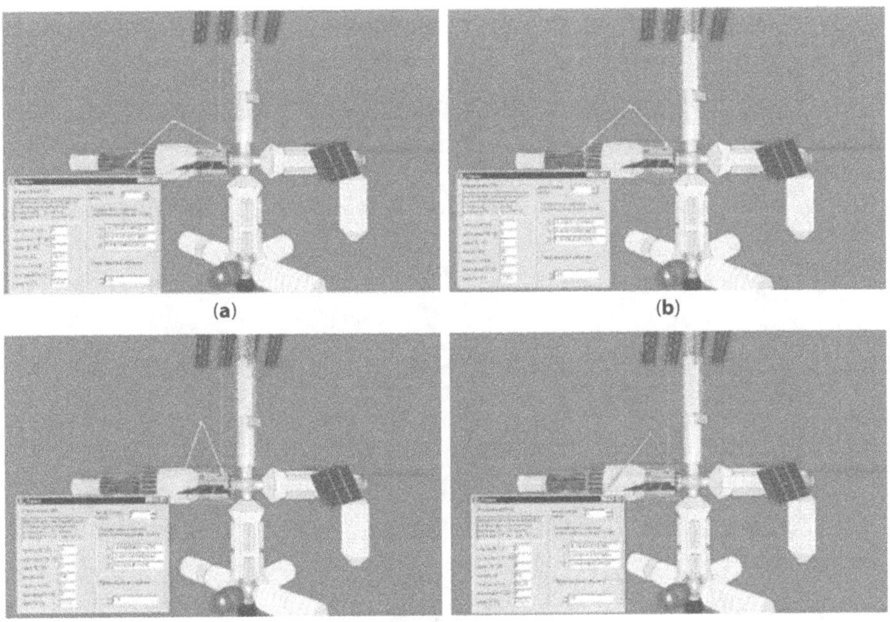

(a) (b) (c) (d)

Fig. 6.39 Frames of interactive motion search.

6.4.4 Experimental Search for a Safe Trajectory of ERA Movement

The search technology of safe EVA, in accordance with the previously adopted algorithm in the TMS environment, contains the following main steps [19].

1. Determination of the current state of the manipulator, the position of all the engines according to the last generated time diagram and state board.
2. Setting the target point using one of the available virtual targeting methods: along the coordinate axes of the local coordinate system in Fig. 6.40 or with the help of an auxiliary spatial coordinate grid.
3. The preliminary formation of a rational trajectory of movement of the working point of the manipulator and the identification of points of collisions (Fig. 6.41). In case of their absence, the manipulator moves in the automatic control mode. The figure shows the trajectory of a yellow intermittent curve. The point of collision is a red sphere. The state panel displays the manipulator parameters corresponding to the collision point. To determine collision points, equidistant cylinders are called for the required ISS model elements.

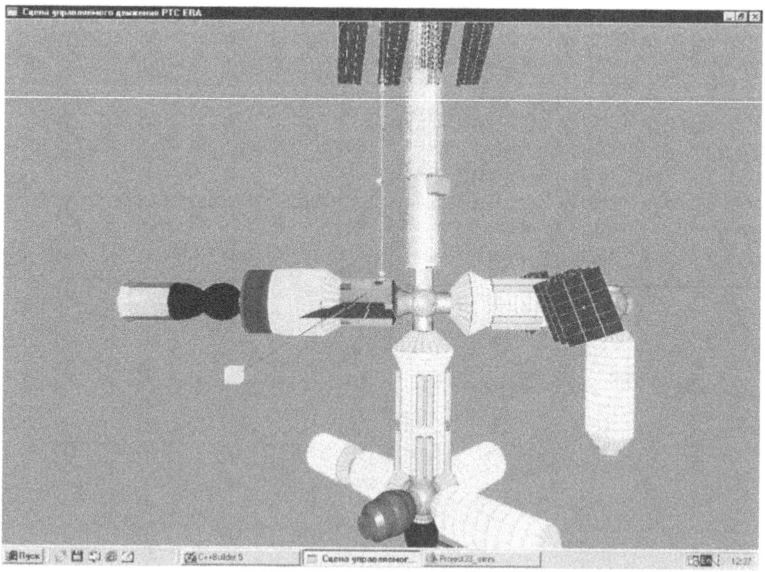

Fig. 6.40 Target point setting.

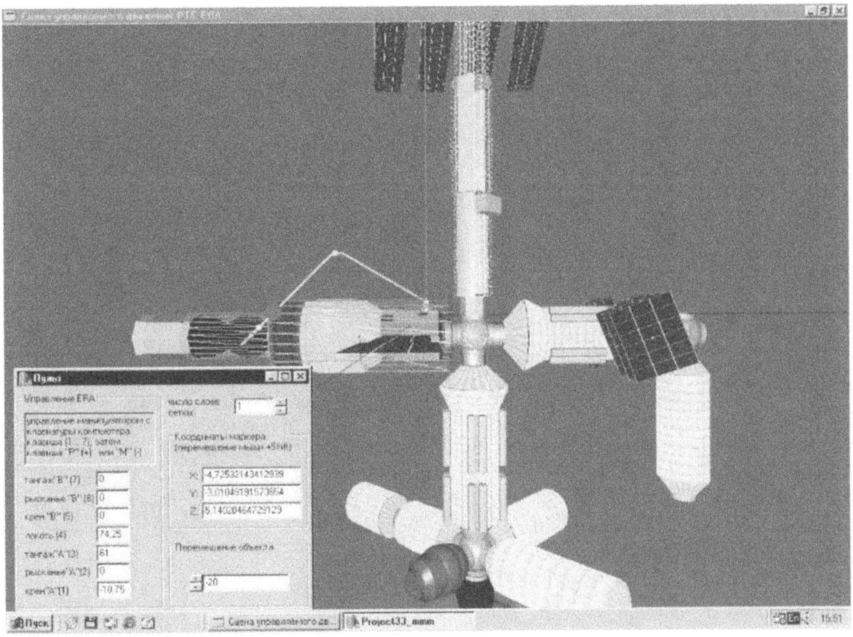

Fig. 6.41 Definition of a rational trajectory.

The manipulator moves to the first point of collision in the automatic control mode (sequential operation of two engines with display of positioning on the display) (Fig. 6.42).

The search for a safe trajectory of the manipulator in manual mode at the point of the second (output) collision or at some valid point on the originally defined trajectory of movement at the end point is presented in Fig. 6.43. The control was carried out using a virtual simulator of the control panel displayed on the working window.

At the same time, an automatic formation of a time diagram (protocol) of engine operation is performed. After finding an acceptable ERA motion path, the manipulator motion program is generated automatically.

6.5 Conclusion

Scientifically, the main hypothesis, which is the basis of the total discretization technology, on the existence of geometrically optimal discretization of curves, was confirmed. The implementation of such a discretization became possible due to the introduction of the concept in the theory of

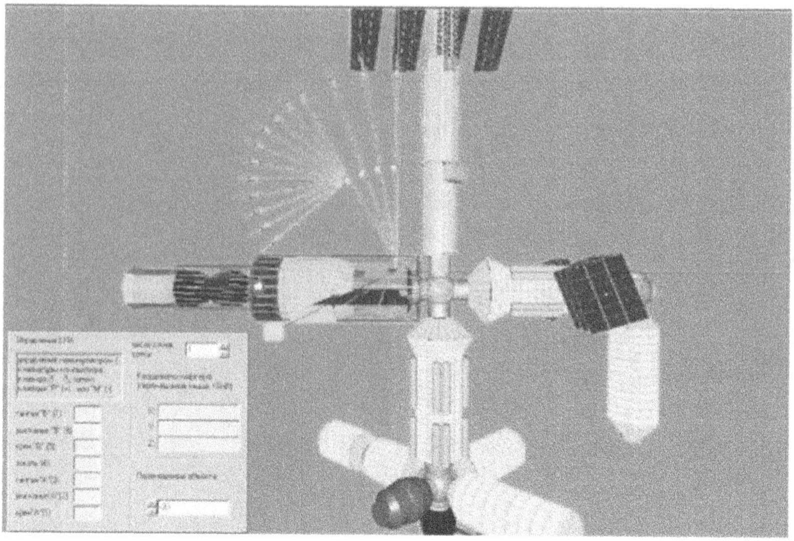

Fig. 6.42 Selection of the motion of the manipulator

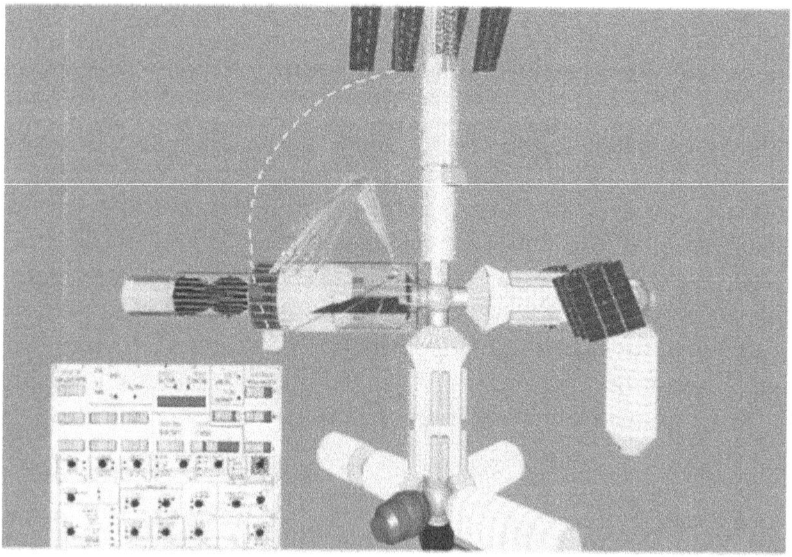

Fig. 6.43 Search for a safe trajectory.

representation of curves – an integral model of a curve. A new approach to solving problems of modeling curved lines and surfaces was substantiated, formulated and implemented, namely, the discrete-integral method, the result of which is a discrete (polygonal) framework.

Even the narrowing of the application domain did not affect the vastness and complexity of the SVR modeling problem with a high degree of realism. The work clearly illustrates the possibilities of geometric modeling as a scientific research apparatus and as a tool for experimental research. Practical confirmation of the effectiveness of the second direction is the developed editor-modeler, which, in addition to the indicated use, has practical value as a software complex for the synthesis and visualization of SVR applied to professional simulators.

In the future, the developed software, such as the virtual designator mechanism, 3D - model development technology, the use of multiple control channels of the RTS, the auxiliary object protection system using the virtual shell, the debugging of the control mechanism on the virtual model, will enable the formation of a complete modeling software environment. This environment will include an editor for setting the RTS operation mode and checking the movement path on a virtual model in animation mode. All this in a complex will allow us to realize the "multi-layered" principle of management organization, when the developed control commands can be refined in the process of animated modeling of the trajectory of the movement.

The subsystems of artificial intelligence also require further development in relation to the subject area under consideration, namely, the creation of self-learning and self-developing simulation training and simulator environments for professional human-machine system operators.

The modern level of development of mechanics and technology allows the creation of a number of complex mechanisms that make it possible to carry out a complex of various operations with unprecedented speed, ease, and efficiency. However, there is a problem of development and safe operation of such systems in very diverse areas of human activity, since managing them is a fairly important matter, and any mistake can lead to irreversible consequences. Thus, the training of personnel for the operation of complex systems, for the practicing complex, abnormal or emergency situations becomes a top priority. The use of simulators is the most optimal way to solve these tasks, but only if they correspond to the modern level of development of graphic software and hardware.

References

1. Lee, V.G., Saprunov, V.N., Kolomiytsev, D.V., Visualization of the training of the extravehicular activity of the RTS cosmonaut-operator in the SVR. *Materials of the 3rd International. scientific and practical seminar "Practice and prospects for the development of institutional partnership." In 2 books*, Edition 1 Publishing house of TSURE, Taganrog, pp. 91–97, 2002.
2. Lee, V.G., Saprunov, V.N., Ulyadurov, A.A., Virtual environment of simulation modeling of training for cosmonauts-operators of robotic facilities of a multi-module space station. *News of TSURE. Thematic issue: Materials All-Russia scientific and technical conf. with the international participation "Computer and Information Technologies in Science, Engineering and Management"*, vol. 1 Publishing house of TSURE, Taganrog, pp. 3–10, 2005.
3. Tao, Y., Xiang, W., Wu, X., SPARC V8 CPU Simulator in Virtualized Verification Systems. Advances in Intelligent Systems Research (AISR), V. 150, 2017. *International Seminar on Artificial Intelligence, Networking and Information Technology (ANIT 2017)*, Atlantis Press, pp. 179–183, 2018. http://creativecommons.org/licenses/by-nc/4.0/
4. Lee, C., Kang, S., Kang, S.G., Kim, K.H., Kim, K., Development of Key Functions for Flight Simulator. *Int. J. Control Autom.*, 9, 1, 347–358, 2016. http://dx.doi.org/10.14257/ijca.2016.9.1.30.
5. Llanos, P., Nguyen, C., Williams, D., Chambers, K.O., Seedhouse, E., Davidson, R., Space Flight Simulator and Mission Control Center: Lessons Learned with XCOR Lynx. *J. Aviat./Aerosp. Educ. Res.*, 27, 2, 2018. https://doi.org/10.15394/jaaer.2018.1736.
6. Zakharevich, V.G., Methods of assessing the activity of a RTS human operator in the TMS environment of virtual reality. *Proceedings of the XIV International Scientific workshop "Practice and prospects for the development of partnership in the field of higher education." In 3 volumes. T.2*, DonNTU, Donetsk, pp. 51–57, 2013.
7. Lee, V.G., Discrete geometric modeling in the problems of the synthesis of objects and processes of the virtual reality environment, 2001.
8. Lee, V.G., Lee, Ulyadurov, *Solution of the collision problem in the virtual workout environment of the RTS cosmonaut-operator*, vol. 9-2, V.G. and A.A. (Eds.), pp. 98–99, News of SFedU. Special Issue: Technical Sciences, SFedU Publishing House, Taganrog, 2006.
9. Lee, V.G., Differential applied geometry in the synthesis of objects and processes of the virtual reality environment, in: *Geometrical computer's model-development. Collection of scientific papers*, vol. 19, V.G. Lee (Ed.), pp. 127–150, Publishing House of the KhSUPT, Kharkiv, 2007.
10. Kosnikov, Yu. N., *Geometric modeling in real-time graphics systems: a monograph*, p. 210 p, Information-edited. Center of PSU, Penza, 2006.

11. Hyatsyntov, A.M. and Mamrosenko K.A., Method of PIP-projections in the visualization subsystem of the training system. *Software Prod. Syst.*, 4, 31–37, 2014.
12. Lee, V.G., *Optimum discretization of curves. Geometric formulation of the problem*, V.G. Lee (Ed.), pp. 8–11, Construction and technological safety. Collection of scientific works of the Academy of Construction and Architecture FGAOU VO "V.I. Vernadsky Crimean Federal University - Simferopol: Publishing house of KFU, 2015.
13. Vygodskiy, M.Ya., *Differential geometry*, p. 511 p, HITTL, M, 1949.
14. Blyashke, V., Differential geometry and geometric foundations of Einstein's theory of relativity, in: *General engineering literature and nomography*, M.-L., Ch. ed., p. 330 p, 1935.
15. Lee, V.G., vol. 1, V.G. Lee (Ed.), News of SFedU. Special Issue: Technical Sciences, pp. 267–268, SFedU publishing house, Taganrog, 2002.
16. Lee, V.G. and Saprunov V.N., Study of the dynamic characteristics of the RTS of the ISS in the environment of the EULER 4 mathematical modeling software package during transportation of payloads//News of TRTU-DonNTU. *Materials of the 5th International Scient. workshop "Practice and prospects for the development of partnership in the field of higher education." In 2 books*, Edition 2, Publishing house of TRTU, Taganrog, pp. 105–114, 2004.
17. Lee, V.G., *Evaluation of the information power of virtual scenes*, vol. 1, V.G. Lee (Ed.), pp. 216–220, News of SFedU. Special Issue: Technical Sciences, SFedU publishing house, Taganrog, 2003.
18. Gromov, G.I., *Differential-geometric navigation method*, p. 384 p, Radio and communication, M, 1986.
19. Lee, V.G., Simulation of collisions when the RTS moves in the virtual space. Virtual modeling, prototyping and industrial design, in: *Materials of the III International Scientific Practical. conf. in 3 vol*, V.G. Lee and V.A. Nemtinova (Eds.), FGBOU VO "TGTU". Tambov: Publishing house FGBOU VO "TGTU", 2016. Vol. 3, vol. II., pp. 127–131, Publishing house FGBOU VO "TGTU", Tambov, 2016.
20. http://www.gctc.ru/main.php?id=221.
21. http://thinkit.ru/blog/viewblog/2249/.
22. http://tft.aero/description.
23. http://www.gctc.ru/main.php?id=221.

Index

www.ingramcontent.com/pod-product-compliance
Lightning Source LLC
LaVergne TN
LVHW020420301125
826312LV00002B/4
* 9 7 8 1 1 1 9 6 6 1 3 1 3 *